Human Variation and Human Microevolution

JANE H. UNDERWOOD

University of Arizona

Human Variation and Human Microevolution

Prentice-Hall, Inc., Englewood Cliffs, N.J. 07632

Library of Congress Cataloging in Publication Data

UNDERWOOD, JANE HAINLINE (date).
Human variation and human microevolution.

Includes bibliographies and index.
1. Human population genetics. 2. Human evolution. 3. Human genetics. I. Title.
GN289.U5 573.2 78-15301
ISBN 0-13-447573-9

Printed in the United States of America

10 9 8 7 6 5 4 3 2

Editorial/production supervision and interior design by Marina Harrison
Cover design by RL Communications
Manufacturing buyers: Gordon Osbourne and John Hall

Prentice-Hall International, Inc., *London*
Prentice-Hall of Australia Pty. Limited, *Sydney*
Prentice-Hall of Canada, Ltd., *Toronto*
Prentice-Hall of India Private Limited, *New Delhi*
Prentice-Hall of Japan, Inc., *Tokyo*
Prentice-Hall of Southeast Asia Pte. Ltd., *Singapore*
Whitehall Books Limited, *Wellington, New Zealand*

FOR MIKE, SAM AND ANNE

Contents

PART 1 INTRODUCTION

PART 2 POPULATION GENETICS

PART 1

Introduction

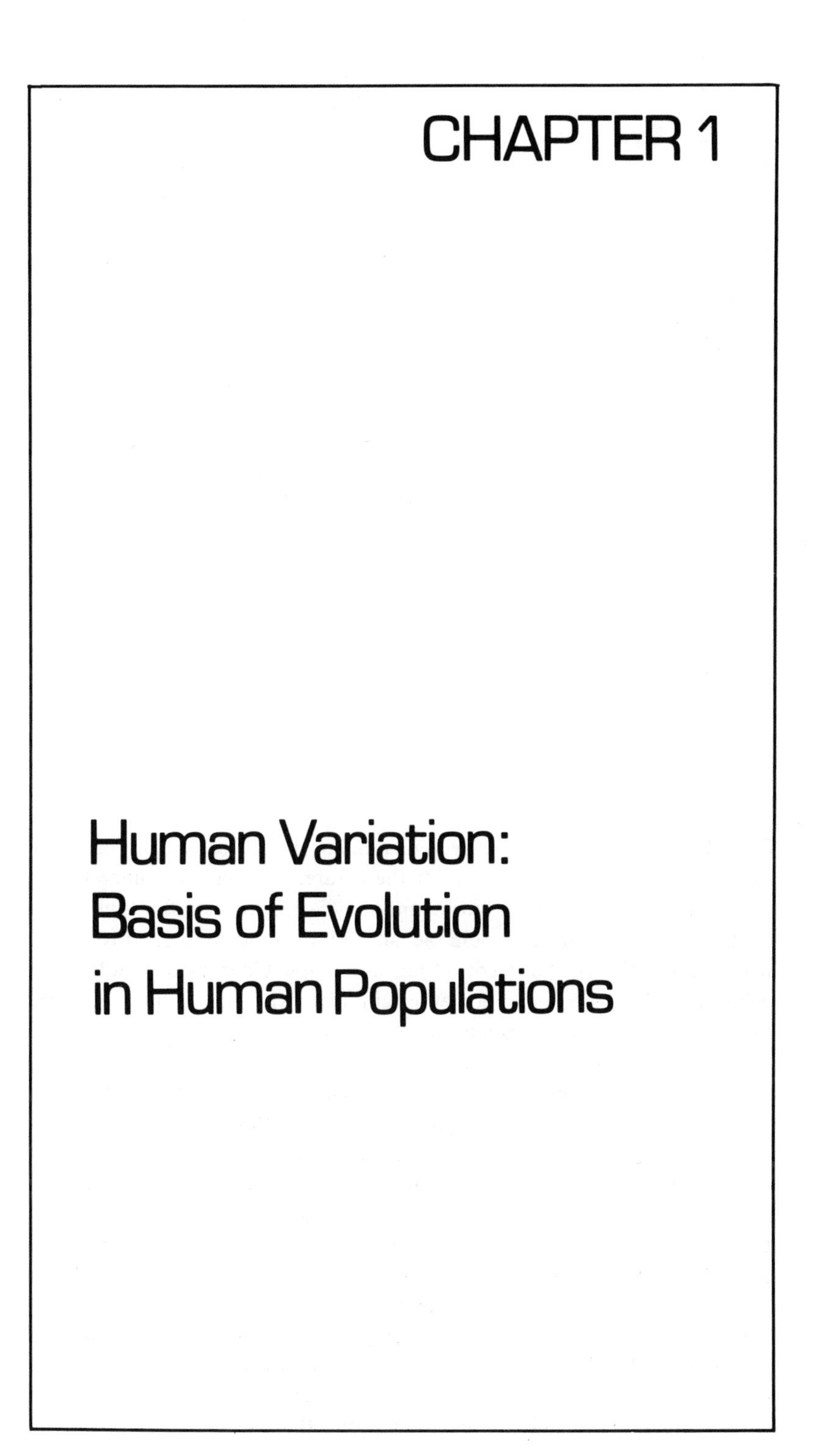

CHAPTER 1

Human Variation: Basis of Evolution in Human Populations

For a long time, people have been examining the creases on the palms of their hands in an effort to predict the future, to reduce the uncertainty of life in an unpredictable world, and maybe just to pass the time when contemplating the navel would be an awkward procedure. Most of us would probably consider consulting a fortune teller who "reads" our palms as a kind of innocuous entertainment, with just a tinge of hope that the more pleasant aspects of the prediction might really come to happen. But how many of us have really looked at the creases, lines, and folds of our hands and compared them to the same features on the hands of our friends and relatives? British authorities in nineteenth-century India did: at least they looked at the lines and grooves on fingertips, and they developed a system for identifying individuals as a measure against impersonation. The use of fingerprints and palmprints for identification is much older than this, of course, and ancient drawings and carvings have even been interpreted as early human recognition of the uniqueness of the individual features of hand prints. Despite the uniqueness of these features, certain patterns of configuration occur, and this has allowed the development of a system for classifying millions of non-identical prints into a limited number of categories. Since the palmar creases are more readily observed without the use of special materials than are fingerprints, it is just as well to begin our consideration of human variation by indulging in a new form of palmistry.

As shown in Figure 1-1, the main palmar creases, however varied, can be classified—in terms of the origin of the three main palmar creases from the radial base crease itself—into one of three categories: single, double, or triple. Many studies have shown that these palmar patterns are under a complex form of genetic direction; that is, the palmar creases on your hands are specified to some extent by hereditary factors passed on to you by your father and mother. Investigations in several human populations have identified significant differences, by sex and by ethnic group, in the incidence of the three patterns.

All of this might remain slightly interesting, perhaps only at the level of passable cocktail party conversation, except that palmar patterns are one of a myriad of possible examples of the incredible range of human variation which most of us never bother to look at, and palmer creases and fingerprints have recently received increasing attention as a diagnostic feature in certain medical conditions. In a sense, palmistry is coming "out of the closet" and is beginning to receive increasing scientific attention. Later on, we will return to the topic of what could be loosely called "medical palmistry"; but for the moment, consider just the implications of all those lines and folds and creases on your hands to a broader appreciation of human diversity.

Each one of us is gloriously, incredibly, wonderfully unique—different from every other human being in the world. Even if you are a member of an identical twin pair, you are not exactly identical in every respect to your twin. And, to steal (and mangle) a phrase, "Vive les differences!" Some of the differences, of course, may be purely a product of the cosmetic industry, but most are the result of the interaction of heredity and environment. We receive genetic instructions from our parents, our biological inheritance, and these directions order our development from a single fertilized egg into a whole functioning organism whose passage from birth through youth and maturity to old age continues to be guided by hereditary information interacting with environmental factors. Identical twins receive identical sets of genetic information, but they inevitably experience at least minutely different environmental conditions; the rest of us receive differing hereditary backgrounds and live under varied environmental settings; so we must all differ to some degree from one another.

We vary in height and weight, in hair color and eye shape, in foot size and hip width, and in so many unseen biochemical traits as to boggle the mind. And so, consciously or unconsciously, we utilize classifications to subdue these innumerable variations into some manageable categories.

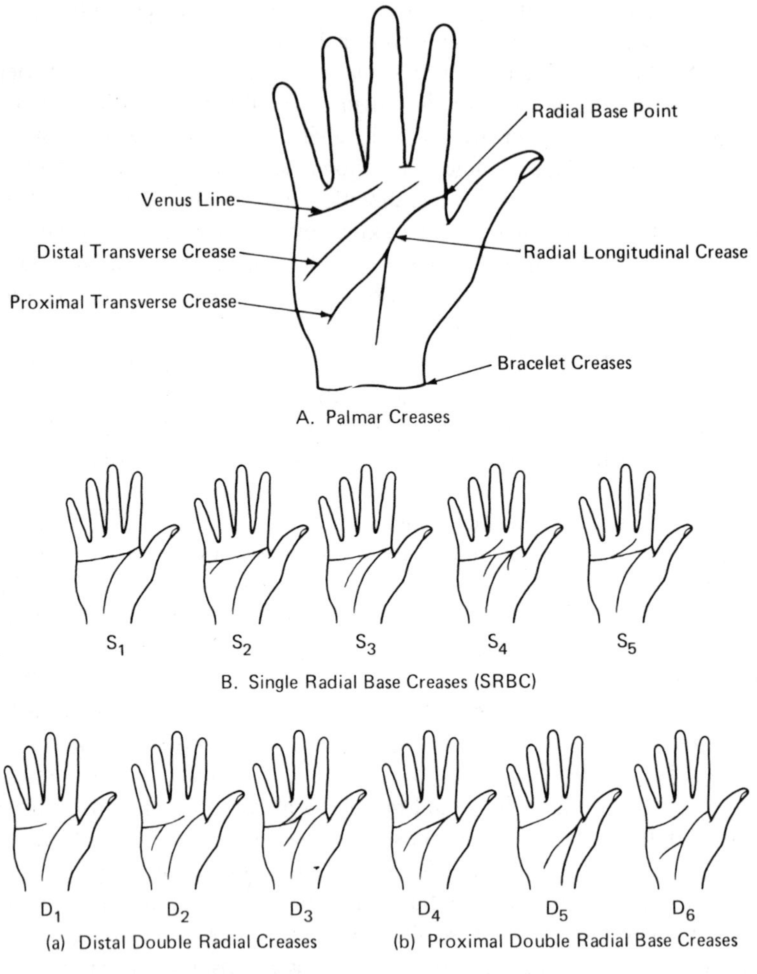

D. Triple Radial Base Creases (TRBC)

FIGURE 1-1 Palmar Crease Pattern Classification (adapted from Chaube, 1977, courtesy of Wistar Institute Press.)

All too often, however, we let the categories we employ assume some kind of false reality, and, except for those individuals with whom we have close relationships, we simply ignore differences. In its most pernicious form, this kind of procedure may become associated with judgments extrapolated from one individual or experience to a whole class of individuals. One or a few disappointments can't provide a reasonable basis for concluding that bankers are lacking in compassion, any more than an examination of several pygmies can justify the conclusion that all individuals with brown eyes are very short.

There are some real limitations on human variation, of course. Palm creases do not appear in arabesque designs, although the added wrinkling and folding of age may increase the complexity of the basic linear pattern. There are no green people walking down the streets, nor any who totally lack the antigens of all the blood types we study in humans. Even if we looked at only ten traits which could each appear in only one of two alternative forms, over 1,000 unique combinations (i.e., $2^{10} = 1024$) would be possible. Now, complicate this further by adding some degree of variation due to external influences, and the permutations could stretch a computer's capacity. While there are limits to human variation, the possible combinations of the thousands of ordinary traits which we examine are simply too great to expect any one of us to have a double, a *Doppelgänger,* running around anywhere in the world.

For many years, scholars tried to examine and to classify this diversity into a system which would at least comprehend all major forms of human variation. Systems were devised to classify people by skin color, since this was a readily perceived difference. Unfortunately, human diversity usually doesn't conform to tidy schemes of organization. "Black" doesn't cover the shades and nuances of skin color in populations which range from olive to dark brown, especially in contrast to "white" populations whose members may also vary, but from slug-white to swarthy olive. Even more complex systems, based on a number of traits, proved unsatisfying because these, too, treated variation as an unpleasant discomfort rather than the essence of the biological reality.

Typological classifications such as the one on the following page (Deniker 1926), as gradually became evident, are really sterile exercises which only begin to comprehend variation when they achieve a degree of complexity equal to that with which they start. In all such efforts, no attempt was made to get at the fundamental processes responsible for the diversity being studied. The real issue is not the ways in which we differ, but why we vary from one another.

There has never been a dearth of explanations of human differences. Chinese writers of the third century B.C. turned to heredity to explain the

Straight-haired peoples
- Asiatic groups
  - Arctic group (e.g., Lapps, Asian circumpolar peoples, etc.)
  - Paleoeans (e.g., Chinese, Japanese)
  - Proto-Malayans (e.g., Southeast Asians)
- Amerind groups
  - Eskimo
  - Paleoindian
  - Variety of Amerind "types"

Wolly-haired peoples
- Eastern group
  - Negritos (Andamanese, Semange, Filipino Aeta, New Guinea Tapiro)
  - Melanesians & Papuans
- Western group
  - Bushmen & Hottentot
- True Negroes - Guinea Coast through Nilotic region
- Pygmies (=Negrillos)

Curly-haired group
- Proto-nordics (e.g., Ainu)
- Proto-indics
  - South Indian jungle tribes
  - Veddoids
  - Sakoi & Senai of Malay Peninsula
- Australian aborigines
- Eastern group
  - Nesiots - S.E. Asia, S. China, Phillippine Islands
  - Chersiots - South India
- Western groups
  - Proto-Egyptian
    - Eurafrican
    - Bedouin
    - Mediterranean
  - Round-Heads (=Eurasiatic)

FIGURE 1-2 Racial Classification (after Deniker *et al.*)

disgusting appearance of the yellow-haired, green-eyed barbarians from distant provinces. Since the Chinese of this period were obviously derived from the matings of human males and females, these aberrant foreigners must have appeared to have derived from a different paternity, perhaps from the breeding of dogs and humans. Aristotle sought psycho-environmental correlates which, not surprisingly, found north Europeans to be a spirited people lacking in intelligence, the Asians an intelligent group lacking in spirit, and the Greeks a spirited *and* intelligent bunch. Leonardo da Vinci even sought a causal explanation for variations in pigmentation: blacks, he argued, living in hot climates, could only work in the cooler temperatures of night, and thus, they became dark in color;

whites, in cooler climates, become lighter by virtue of their daytime activities. The Cherokee Indians of North America turned to a fallible Creator in their search for an explanation of differences in skin color, and concluded that the Creator had formed paleskins by underbaking his product, blacks by overbaking, and Indians when he had perfected his techniques.

These and a myriad of other pre-Darwinian explanations of human variation embody a typological approach that envisions variation as the imperfect reflection of a more real essence. Modern appreciation of human variation and our understanding of the processes underlying it rest on *population thinking,* what Mayr (1977) has called "one of the most drastic conceptual revolutions in Western thought," a concept stressing the individuality of organisms. Charles Darwin's ability to incorporate the idea into his thinking, and to combine it with his observations as a naturalist and the products of his reading, especially of Malthus' essay on population growth, established the foundation of modern evolutionary theory.

Charles Darwin, whose theory of the origin of species through natural selection revolutionized our understanding of human evolution, ought to be the patron saint of college students. Trained in medicine (his father's profession) at Edinburgh, he felt unsuited to the profession and moved to Cambridge with the notion of entering the ministry. But, after a summer's field trip to Wales in company with a geologist, he was offered an unpaid post as naturalist aboard a forthcoming survey voyage around the world on the ship *H.M.S. Beagle.* Actually, Darwin was the third choice for the post. As it turned out, he scarcely qualified as an ardent seaman, for he suffered from seasickness repeatedly for the duration of the trip which lasted from 1831 to 1836. Darwin's journal was published as one of several volumes comprising the official report of the journey, but his theory of evolution was not published until 1859, a delay which should make the most tardy writers of term papers feel somewhat elated. In fact, Alfred Wallace had independently reached similar conclusions and developed a remarkably similar theory of evolution, about which he wrote to Darwin in 1858; only then did Darwin hasten to complete his writing, producing a single volume exposition of his views and ideas.

Given that the members of all animal species differ slightly from one another, Darwin saw that gradual changes would occur if some of these variants survived to produce relatively more offspring than others. If this occurred over a number of generations, then a modern species might differ greatly from its ancient ancestors—not as a result of some cataclysmic event, but as a result of the continuous shift of the range of variation characterizing a species at some distant time in the past toward the range of variation found in the modern form. Obviously, this view also implied that modern forms were subject to the same process, that members of a species at some future time would represent a different range of variation than could be observed at the present. The postulated mechanism for these changes is best described in Darwin's own words:

> As many more individuals of each species are born than can possibly survive; and as, consequently, there is a frequently recurring struggle for existence, it follows that any being, if it vary however slightly in any manner profitable to itself, under the complex and sometimes varying conditions of life, will have a better chance of surviving, and thus be *naturally selected.* From the strong principle of inheritance, any selected variety will tend to propagate its new and modified form.[1]

But Darwin, along with other scholars of that period, knew little of the mechanisms of genetic inheritance and felt that some kind of merging of hereditary traits must take place. This posed a serious problem for the acceptance of Darwinism, for if offspring are a meld of parental variations, how can natural selection of variant forms work? Six years after the publication of Darwin's great work, an Austrian monk, Gregor Mendel, reported the results of his laborious studies and analyses of breeding experiements with pea plants. Mendel was able to show, by examining a few traits (color, shape, height), that the progeny of distinctive parents were not a blend of parental attributes.

When short plants were crossed with tall plants, the offspring were not of medium height, as the blending notion would have required, but all progeny in the first filial (F_1), or offspring, generation were tall. Crosses between individual plants of the F_1 generation reproduced two kinds of offspring, tall or short, in a ratio of about three to one in the second (F_2) generation. Mendel argued that these results could only be explained by assuming that the parental generation was composed of two kinds of genetically different forms: tall plants, which passed on a single particle of genetic information coding for tallness in the offspring; and short plants which also passed on genetic information of a discrete nature, but coding for short height. In the F_1 generation, each individual plant would have received both units, one from each parent; but only one of the units, that for tallness, was expressed in the external appearance of the individual plant. Yet, each unit must have remained discrete, since in the F_2 generation, some short plants appeared as well as many more tall ones. In the light of our present understanding of genetic inheritance, we would have described the genetic constitution of the parental (*P*) genera-

[1] Darwin 1958, Mentor, p. 29.

tion as being either *TT* or *tt*, and that of the F_1 generation as being *Tt*. When individuals of the F_1 generation transmit genetic information to the offspring (F_2) generation, they can pass on either the *T* unit or the *t* unit, and these can combine with the *T* or *t* units formed by other F_1 plants to form three possible combinations of genetic information in the F_2 offspring: *TT*, *Tt*, or *tt*. Some of these new plants will be tall (*TT* or *Tt*), but a lesser number will be short in appearance (*tt*). Clearly, the material of heredity must behave as particulate units to attain these observed results.

Mendel's work remained unread, not only by Darwin but by most scientists, until around the turn of the century. Even after its rediscovery, many questions still had to be resolved. Studies in a wide range of biological populations, not least among which were anthropologists' reports on the lifeways and cultures of human societies around the world, showed that the notion of a fierce, bloody, internecine, competitive struggle for limited resources was not the inevitable accompaniment of evolution. Morgan and his co-workers had already begun their extended studies of the common fruit fly, *Drosophila melanogaster*, and demonstrated that most variations involved slight differences, both in external form and in

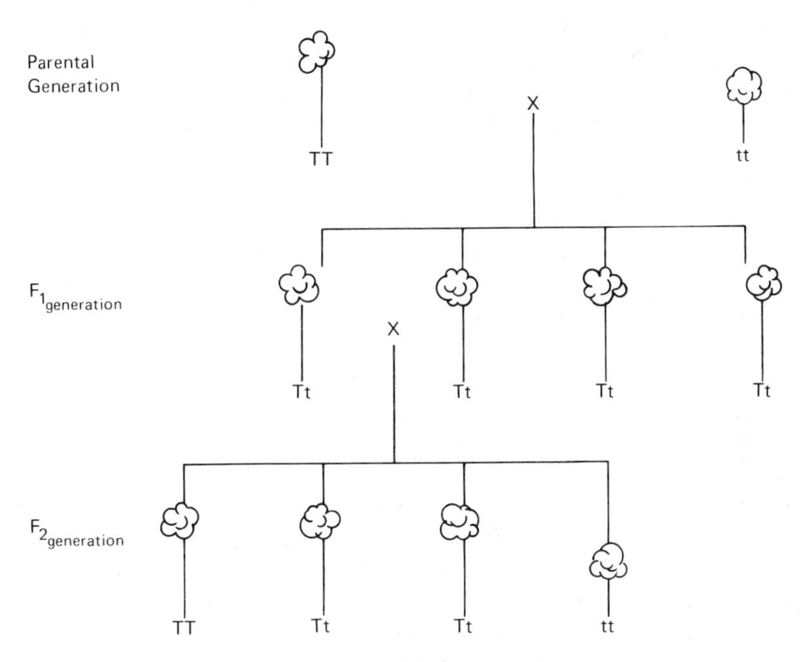

FIGURE 1-3 Diagram of Mendelian Inheritance

the hereditary material. The possessors of favored differences did not simply annihilate all other members of the population, but usually passed on their genetic material only slightly more frequently than did their less fortunate peers. This differential in reproductive success, perhaps of the order of only 1 or 2 percent, could effect as great a change over a number of generations as could a violent struggle to the death between competing forms.

By the 1930s the work of the fruitfly geneticists, of biometricians such as Haldane, Galton, and Fisher, and the contributions of an untold number of naturalists, botanists, and other biologists provided the basis for the new synthesis of evolutionary theory, as expressed in the writings of such men as Ernst Mayr, Theodosius Dobzhansky, and others. The essence of their work was an emphasis on evolution as a gradual process that involved changes over time in the composition of the genetic properties of biological populations. Natural selection, the differential contribution of carriers of certain genetic combinations, provides the primary mechanism of this process. Mutations, changes in the genetic material, introduce new hereditary materials into the population; the exchange of genetic materials between populations can provide new combinations, and random fluctuations can contribute unique contrasts among different populations on which natural selection can also operate.

As we now know, natural selection does not operate exclusively through differential fertility or through differential mortality, but always operates so as to result in the differential contribution of genetic materials from some individuals to the genetic composition of the population in the next generation. A child born with a serious genetic disease, such as cystic fibrosis, rarely survives long enough to reproduce, and thus makes no contribution to the next generation. Yet, if this is so, the incidence of cystic fibrosis should be much lower than it is. Some studies suggest that the parents of children suffering from cystic fibrosis have more *siblings* (brothers and sisters) than do parents of unaffected children. As in the case of Mendel's pea plants, the parents of children afflicted with this disease must themselves be carrying the genetic information (*c*) which produces the disease in offspring (*cc*) who receive a genetic unit for the condition from both parents. It follows, then, that although the stricken children die before transmitting this genetic inheritance (differential mortality), the genetic unit responsible for the condition has been maintained in the population through the differential reproduction of carrier (*Cc*) parents. A more thorough description of the complex operation of natural selection in human populations is found in Chapter 9. This brief introduction provides occasion to return to our initial consideration of the incredible variation found in all human groups.

The mechanisms of Mendelian inheritance are identical in all human populations. Whatever genetic information is transmitted from parent to offspring is carried by the *chromosomes,* stainable bodies visible in the nucleus of the sex cells which parents produce and contribute to form the new organism. As we receive half our chromosomes from one parent and half from the other parent, we in turn pass on only one half of the entire number of chromosomes we have received from our parents to each of our children. This process can be illustrated by looking at only one pair of the chromosomes, the sex, or *X* and *Y*, chromosomes which play a critical role in sex determination.

Most of us have one of two possible combinations of this pair of chromosomes present in the nuclei of most of the cells of our bodies. Either both chromosomes are similar (*homologous*) in form and appearance, the *XX* combination, or one *X* chromosome is accompanied by a much smaller, *Y*, chromosome, the *XY* constitution. Individuals with the *XX* pair have received one *X* from the father and one *X* from the mother, and normally have the external appearance and internal sexual organs (ovaries, uterus) of a female. Those with the *XY* constitution usually have male sexual organs (testes, penis) and appearance, and have received an *X* chromosome from the mother and a *Y* chromosome from the father.

The study of the sex chromosomes is greatly facilitated by the *sex chromatin, or Barr body, test.* When properly stained, cells taken from a normal *XX* female are found to have a single small structure, the Barr body, present on each nucleus, but this structure is lacking in cells taken from *XY* males. The number of Barr bodies present is always one less than the total number of *X* chromosomes present. Mary Lyons has suggested that the Barr body is actually a genetically inactivated *X* chromosome, and this theory would help to explain the similarities in expression of certain traits in females with their two *X* chromosomes and in males who have only a single *X* chromosome. The availability of this relatively simple laboratory procedure for diagnosis of chromosomal sex has made it possible to identify a number of sex chromosome variants in many human populations.

Occasionally, individuals are found who have either an excess or a deficiency of sex chromosomes. In the condition known as *Klinefelter's syndrome,* infertile humans with the external genitalia of males are found to have two, three, or even more, *X* chromosomes in addition to a *Y* chromosome. The discovery of such individuals suggests that the presence of the *Y* chromosome is the critical factor that directs the development of embryonic organs into the pattern characteristic of males. In some other species, the sexual development of the male is dependent on the relative number of sex chromosomes; in human females the rare *XXX* chromosome combination is usually found in normal-appearing women; males with the *XYY* combination are often reported to be tall (over six feet in stature), fertile, but with normal sexual organs and genitalia.

The absence of one of the sex chromosomes has far more serious consequences than the presence of one or more extra chromosomes. No *YO* individual has yet been identified, which suggests that the development of an embryo with this combination is so seriously affected as to preclude survival. However, individuals with an *XO* constitution (*Turner's syndrome*) do occur. Externally, the Turner's syndrome has a generally female appearance, but the ovaries are rudimentary or absent; such individuals are of short stature, and show underdevelopment of the secondary sexual characteristics. From these facts, we can only conclude that the presence of at least one *X* chromosome is necessary for the development of a viable male or female, that supernumerary *X* chromosomes probably become inactivated Barr bodies and have minimal to moderate effects on development in males, while the presence of a *Y* chromosome is essential for the development of male characteristics.

Although these principles of sex determination apply to the species as a whole, human populations do vary in a number of traits related to sex determination. The average length of the *Y* chromosome differs among populations, being greater in Japanese and smaller in non-Jewish whites. The frequency of several of the sex chromosome abnormalities is far from uniform in all human groups. But differences are most readily seen in comparisons of the sex ratios in various human populations.

The basic mechanism of sex determination, as described, should result in equal numbers of male or female conceptions, or a *primary sex ratio* of 100. Since conceptions can scarcely be directly observed, we have to look first at the *secondary sex ratio,* the relative number of males and females at birth. In the United States, this figure varies from about 106 (that is, 106 boys are born for every 100 female births) in whites to slightly less than 103 in American blacks. Although great variance is found in the sex ratios from small populations, and may be ascribed to chance variations, studies involving tens of thousands, even millions, of births have shown values of from about 110 to 95 in the sex ratio of different populations.

If the primary sex ratio is equal, but the secondary sex ratio in United States whites is skewed in favor of males, then differential mortality of female embryos and fetuses must be taking place. Yet, available data from the United States Census Bureau reports on stillbirths show a consistent excess of males, and a classic morphological and histological analysis of over 6,000 embryos and fetuses in the Department of Embryology at Carnegie Institution found an excess of males among abortuses.

If these data are accurate, then the primary sex ratio must be grossly skewed in order for an excess of males to be present at birth.

With the advent of more sophisticated techniques of chromosome analysis of fetuses, more reliable diagnoses of sex have become available, and these do not indicate a gross excess of deaths in male embryos. It is still impossible to determine the sex ratio at conception. But if equal numbers of males and females are conceived, more female embryos must die before birth to explain the observed secondary sex ratio; there is now no convincing evidence that this does in fact occur.

An alternative explanation would require that more male than female conceptions take place and that this is reflected in the secondary sex ratio. Again, evidence in support of this idea must be indirect. Since the sex of the offspring is determined solely by the father (who contributes either the *X* or *Y* chromosome), differences in *Y*-bearing sperm, either in speed of access to the female's egg, or in ability to penetrate the egg, could account for a primary sex ratio greater than 100. In fact, claims have been made that *X*-bearing sperm can be distinguished from *Y*-bearing sperm in terms of mass and of swimming ability. *Y*-bearing sperm carry a shorter sex chromosome, which has less genetic material, and thus have less mass than an *X*-bearing sperm. The lighter sperm have greater motility and, all other factors being equal, should be able to reach the ovum somewhat in advance of *X*-bearing sperm. Once the sperm nucleus penetrates the ovum, the female sex cell is usually impervious to penetration by additional sperm, so the early arriving sperm may well have an increased advantage over the heavier, slower *X*-bearing sperm.

But all other factors are probably *not* equal in sex determination, and many students of the subject have argued that a causal connection exists between the time of fertilization within the menstrual cycle and the sex of the conceptus. Fertilization in the earlier part of the menstrual cycle results in a higher sex ratio, so James (1976) has argued that the rate of sexual intercourse plays a large role in this matter.

One of the interesting observations in human sex-ratio trends is the increase in sex ratio of newborn infants in belligerent countries during and just after wars. This, it is argued, results from the high rates of sexual intercourse for servicemen home on brief leaves and during early periods after veterans are demobilized. This argument is further supported by a number of studies which indicate, directly or indirectly, that frequency of coitus is positively associated with masculinity of the sex ratio. For example, the sex ratio is lower for Jewish than non-Jewish births in Israel. This has been related to the Jewish Orthodox ritual of Niddah, the practice of sexual abstinence for a period of one week after menstrual bleeding has ceased. Thus, while the hereditary mechanism of sex determination is uniform among human populations, even cultural practices affecting sexual behavior may influence the expression of this trait as seen in the observed variations in the secondary sex ratio.

Individual variation, then, can only be understood in terms of the interplay between the genetic instructions, or *genotype,* and the multitude of factors which can affect that trait's expression, or the *phenotype,* of an individual. Two *XX* females within one population may differ greatly from one another in the expression of such secondary sexual characteristics as body hair distribution or breast development, but both are functioning, normal females. In contrast to normal *XX* females from a second population, even greater differences in phenotype may be noticed. There is, in short, no female "type," but a range of variation in expression of phenotypes characteristic of different populations, and some overlapping of these population ranges does occur. Merchants can expect a more limited demand for larger brassiere sizes in Thailand's stores than in the merchandise centers of the Midwestern United States, but woe betide the store owner in either part of the world who fails to stock the entire range of available sizes.

Some variations are self-limiting, since they are not passed on to offspring and reappear as a result of errors in the formation of functioning sex cells in unaffected adults. Turner's syndrome females and Klinefelter's syndrome males who lack functioning sexual organs cannot transmit this chromosomal disorder to children. In the vast majority of reported cases, fertile *XXX* women have produced only normal *XX* or *XY* children, suggesting that any *XX* sex cells formed are not viable.

Many genetic traits have less drastic effects on the phenotype than do those involving whole sex chromosome combinations. Some of these, too, are self-limiting because affected individuals die before they begin reproductive activities. Thus, despite medical advances, victims of cystic fibrosis disease rarely reach adulthood. In other genetic conditions, such as one form of dwarfism, chondrodystrophic dwarfism, it has been claimed by some authorities that affected adults simply have fewer children than do their unaffected brothers and sisters. But what about all those seemingly minor genetic traits which appear to be relatively insignificant or just inconsequential to the changes of an individual surviving and producing children?

In fact, there may well be traits with neutral effects on survival and reproduction, but a number of these presumably neutral traits have been associated, or correlated, with conditions which have a bearing on survival and the probability of passing on genetic information to offspring. Significant associations have been reported between patterns of palmar creases and such disorders as leprosy, tuberculosis, schizophrenia, and

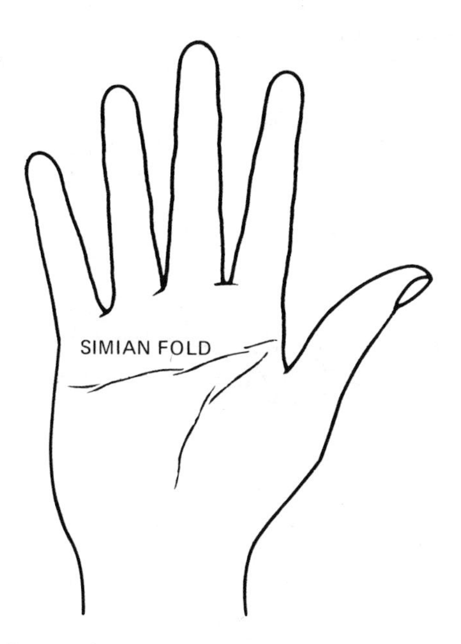

FIGURE 1-4 Simian Fold

diabetes mellitus (Eswaraiah and Bali, 1977), thus bringing palmistry to the attention of the medical world. Forty percent of children suffering from Down's syndrome (formerly known as Mongoloid idiocy), associated with certain kinds of chromosomal abnormalities, have a characteristic palmar crease known as the simian fold (Figure 1-4), whereas this trait is found in only some 10 percent of the population as a whole. Investigations are underway which seek to trace an apparent association between certain fingerprint patterns and defects in the mitral valve of the heart.

Correlations do not prove causality and must be regarded with caution, but associations of this kind suggest testable hypotheses which have often led to a greater understanding of how human variation is maintained. It has only been possible in this brief introduction to sketch some of the questions which anthropologists and human biologists are investigating and to suggest some notion of the kinds of theoretical and methodological concerns of these studies. Above all, our efforts begin with the recognition and appreciation of human diversity.

REFERENCES AND RECOMMENDED READINGS

CHAUBE, R. 1977. Palmar creases in population studies. *American Journal of Physical Anthropology* 47:7–10.

DARWIN, CHARLES 1958. *The Origin of Species by Means of Natural Selection or the Preservation of Favoured Races in the Struggle for Life.* (Mentor edition). New York: New American Library of World Literature.

ENGEL, L., ed. 1962. *The Voyage of the Beagle,* by Charles Darwin. Natural History Library edition. Garden City: Doubleday Anchor Book Company.

ESWARAIAH, G., and R.S. BALI 1977. Palmar flexion creases and dermatoglyphics among diabetic patients. *American Journal of Physical Anthropology* 47:11–14.

JAMES, W.H. 1976. Timing of fertilization and sex ratio of offspring–a review. *Annals of Human Biology* 3:549–556.

MAYR, E. 1977. Darwin and natural selection. *American Scientist* 65: 321–327.

CHAPTER 2

The Genetic Basis of Human Diversity

However varied the forms and features of humanity, we all begin in the same way—from the creation of a *zygote,* a fertilized egg cell produced by the fusion of male and female *gametes,* the spermatazoon and the ovum. These gametes contain the entire body of genetic information passed on to us by our parents, and the gametes we form in turn and transmit to our offspring contain all the genetic material which our children will receive from us. Most, if not all, of that information is carried in the chromosomes which are present in the nucleus of each reproductive cell, and these cells are the packages of genetic information and accessory materials upon which the survival of our species depends. Clearly, no biological process is more fundamental to the study of human evolution than that of *meiosis,* the special kind of cell division which results in the formation of eggs and sperm.

Most of the cells in our body result from the multiplication of the original cell, or zygote, through a process called *mitosis.* This process involves: the replication of all 46 chromosomes (the *diploid,* or 2N, chromosome number) contained in the parental cell; the subsequent separation, or division, of each duplicated chromosome; and the formation of new nuclear membranes enclosing each nucleus with its complement of 46 duplicated, or daughter, chromosomes. This complex sequence of events ensures that new cells contain replicas of all the genetic materials contained in the ancestral cell. Were this same process to be followed in the formation of the gametes, however, the later fusion of two such cells during fertilization would mean a doubling of the genetic materials—from 46 to 92 chromosomes in the first generation, from 92 to 184 chromosomes in the next generation, and so on. In *gametogenesis,* the formation of sex cells, meiosis results in the production of *haploid* (N chromosome number) gametes containing only 23 chromosomes. The fusion of two gametes in fertilization restores the basic chromosomal complement of 46 chormosomes. An examination of Figure 2-1, a simplified illustration and comparison of meiosis and mitosis, illustrates the main features of each process and indicates several of the significant differences between them.

One of the more critical distinctions between the two processes occurs during late prophase I and early metaphase I when, in meiosis, but not in mitosis, the homologous chromosomes, now in duplicated form, align with one another along the spindle fibers. At this time, breakage of the *chromatids* (the duplicated chromosomal strands) may occur, and crossing over of portions of the sister chromatids may take place. Consider, for example, what might ensue if the diploid precursor of the gametes to be formed contained a maternally derived chromosome which contained genetic instructions *A* and *B,* whereas the paternally derived chromosome carried the genetic instructions *a* and *b.* When the duplicated chromatids align, this information could be represented as in Figure 2-2. If no crossing over occurs, only two kinds of gametes can be formed—those with the *AB* sequence or others carrying the *ab* instructions. Breakage and crossing over, however, can result in increased variation through the formation of gametes with the *AB, ab, Ab,* or *aB* constitution. This *gene recombination* provides a rich source of genetic variation independent of the need for any actual change in the genetic materials, or mutation.

In humans, spermatogenesis takes place throughout the adult lifespan of the individual. As a boy reaches adolescence, the pituitary gland begins to release several hormones which stimulate the secretion of male hormones and the production of *spermatagonia,* the diploid germ cells from which the mature germ cell, or spermatazoon, will develop. The spermatagonium, through mitosis, produces diploid cells, some of which become diploid *spermatocytes.* These undergo meiotic division to produce haploid *spermatids,* which in turn undergo a number of changes, forming the characteristic structure of a mature *spermatazoon,* complete with a head, midpiece, and tail. The head of the sperm consists mainly of the nucleus—which now includes the haploid number of chromosomes, and the acrosome—a structure containing several enzymes.

Shortly before ejaculation, several accessory glands produce secre-

MITOSIS

PROPHASE
Chromatids present

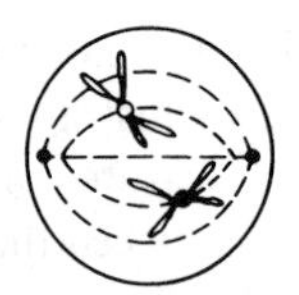

METAPHASE
Centromeres align independently on spindle

ANAPHASE
Centromeres divide, new chromosomes move apart

TELOPHASE

Nuclear membrances form around two new nuclei, each containing 2 N number of chromosomes

MEIOSIS

PROPHASE I
Chromatids present; homologous chromosomes align and crossing over may occur

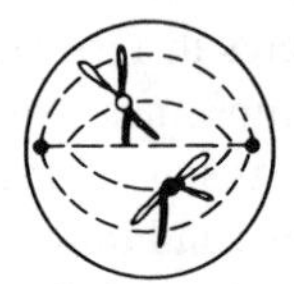

METAPHASE I
Centromeres align in pairs on spindle

ANAPHASE I
Centromeres do not split, but move apart

TELOPHASE I

Two new nuclear regions form, each containing one centromere but duplicated chromatids

PROPHASE II
Each cell contains one centromere with duplicated chromatids

METAPHASE II
Centromeres move onto new spindles

ANAPHASE II
Centromeres divide, new chromosomes move to opposite poles

TELOPHASE II
Four cells form, each with haploid (N) chromosome number

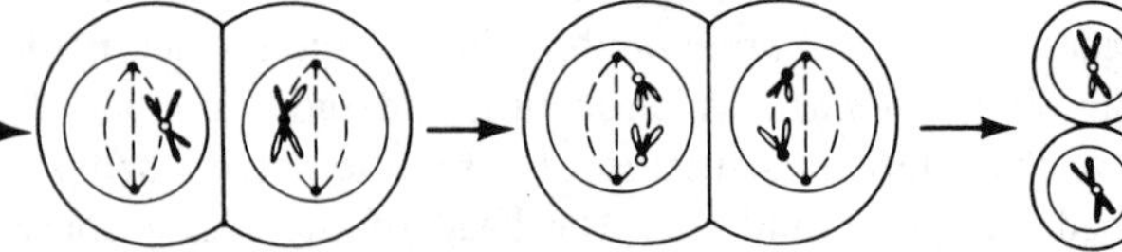

FIGURE 2-1 Simplified Diagram: Comparison of Meiosis and Mitosis (only one chromosome pair shown)

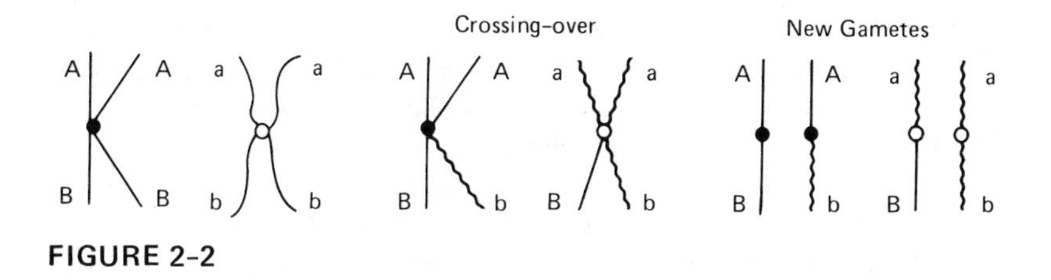

FIGURE 2-2

tions which, together with spermatazoa, form the semen which will be deposited in the vaginal tract during intercourse. As the seminal fluid thins after deposition in the vagina, the sperm become motile and begin moving through the cervix and into the uterus. Some few hundred of the millions of sperm present in the ejaculate may reach the area of the female sex cell as it moves from the ovary into the upper portion of the Fallopian tube. For fertilization to take place, the sperm must penetrate the cells of the corona radiata and the zona pellucida surrounding the egg within about twenty-four hours after the release of the female gamete from the ovary.

Unlike males, who will produce countless millions of gametes throughout the entire span of post-adolescent life, females are born with only a limited number of several hundred thousand oocytes, of which only a few hundred will be released as viable ova during only part of the adult lifespan of most women. At about the time of ovulation, roughly halfway in the menstrual cycle, a *primary oocyte,* under hormonal influence, completes its first meiotic division, producing a large *secondary oocyte* and a smaller, nonfunctional, *polar body* which is extruded. The secondary oocyte is then released and moves into the Fallopian tube. If the female germ cell is successfully penetrated by a sperm while it is moving down the upper portion of the Fallopian tube, the cell completes a second meiotic division and expels a second polar body. The haploid nuclei of the remaining *ovum* and of the sperm now lose their nuclear membranes and merge, thus effecting fertilization. In the next few days, the zygote begins to divide mitotically, until the *blastocyst* (a structured group of cells which give rise to the organism and to the membranes surrounding it) becomes implanted in the endometrial layer of the uterus some five to seven days after fertilization takes place. The *embryo,* as it is now called, begins a series of complex, finely timed developments, which, if successful, culminate in the delivery of a full-term viable infant about 266 days after fertilization occurs.

Despite these differences in spermatogenesis and oogenesis, the formation of mature sex cells in each sex ensures the production of gametes which contain a random assortment of the maternal and paternal chromosomes transmitted from the male parent and from the female parent to the newly formed zygote. The bits of genetic information which each parent passes on to the offspring do not somehow mix or melt into each other, but are transmitted as particulate units, are carried on the chromosomes, and are randomly distributed through meiosis into the two sex cells. The meiotic process makes the findings of the study of human kindreds consistent with the basic principles of Mendelian genetics.

Mendel, who knew nothing of chromosomes or chromosomal behavior, was able to investigate the inheritance of certain traits in peas by a series of carefully controlled breeding experiments. Laboratory breeding experiments are not available to the student of human genetics, but we have access to other valuable sources of information—historical records and genealogies. Long before Mendel's classic experiments were performed, simple patterns of inheritance for a few traits in humans had already been worked out through observation of human pedigrees. For example, the Talmud includes a proviso for exempting from circumcision newborn males whose elder brothers or mothers' brothers showed a tendency for excessive bleeding, implying an elementary understanding by ancient Jewish scholars of the basic pattern of inheritance of one form of hemophilia.

Some traits in humans are inherited as autosomal dominants; that is to say, the genetic instructions for such traits are carried on one of the twenty-two pairs of chromosomes which, unlike the X and Y chromosomes, are not directly involved in the determination of the sex of the individual. These traits are expressed in the phenotype whether identical (AA) or nonidentical (Aa) genetic instructions are present at the same position, or *locus,* on both chromosomes of the homologous chromosomal pair. These coding instructions at a single locus on both members of a chromosomal pair or homologous chromosomes are referred to as *alleles.* For such traits to be diagnosed as the product of autosomal dominant inheritance, we would expect to find in a collection of pedigrees that (barring the rare event of mutation) all affected offspring must have at least one affected parent. In a large group of such pedigrees, we would expect to find that the affected parents may be either males or females, and that both male and female offspring may have the trait.

The basis for these features of autosomal dominant inheritance are readily apparent from a consideration of the behavior of chromosomes during meiosis. We can examine this by looking at one such trait, multiple lipoma—the presence of nonmalignant fatty tumors, usually occurring beneath the skin, and probably the most common form of tumor in adults. An affected parent may have the allele (L) responsible for the trait present at the appropriate locus on both members of the autosomal chromosome pair, in which case the genetic constitution, or genotype, at this locus is said to be *homozygous* (*LL*). Alternatively, the affected parent may have a *heterozygous* genotype, with unlike alleles at the locus (*Ll*), since the presence of the single dominant allele is sufficient to direct the expression of the phenotype for these fatty tumors. Suppose a homozygote (*LL*) individual, male or female, marries an unaffected person (*ll*). All children from such a mating will exhibit the trait in adulthood. During meiosis the homozygous affected parent will produce gametes, all of which carry the *L* allele, while the other parent will produce only *l*-carrying gametes. Thus, the zygote can only receive an *L* from one parent and an *l* from the other parent and must, therefore, have the genotype *Ll.*

In some cases, however, the affected adult will be a heterozygote at this locus, with the genotype of *Ll.* Gametes formed by the heterozygote may carry either the *L* allele or the *l* allele. If a heterozygote mates with an unaffected person (genotype *ll*), two kinds of zygotes may be formed—some with the genotype *Ll* and the lipoma phenotype, and others whose genotype is *ll,* and these latter progeny will be unaffected. If the two kinds of gametes formed by the affected parent are produced in about equal frequencies, the probability of producing an affected offspring is about 50-50. Figure 2-3 may help to illustrate this kind of mating and the combinations which can be produced from it.

In the event of the mating of two affected heterozygotes, L1 × L1, resultant possible combinations can be similarly represented, as in Figure 2-4. In other words, the ratio of possible genotype combinations is 1 LL: 2Ll: 1ll, but the ratio of possible phenotypes is 3 affected: 1 unaffected.

Many traits in humans involve autosomal dominant inheritance, but much of our information on human genetics has come from the study of

Affected Parent, Genotype L1, can produce gametes with

Unaffected Parent, Genotype 11, can produce gamets with		L	1
	1	L1	11
	1	L1	11

Offspring Genotypes:

Ratio of Expected Offspring:
L1 (affected) to 11 (unaffected) = 50:50 or 1:1

FIGURE 2-3 Diagram of Autosomal Dominant Inheritance (L1 x 11 mating)

Parent, Genotype L1,
can produce gametes with either

Parent, Genotype L1,
can produce gametes
with either

	L	1
L	LL	L1
1	L1	11

Offspring

Ratio of Expected Offspring:
LL (affected 1 : L1 (affected) 2 : 11 (unaffected) 1

FIGURE 2–4 Diagram of Autosomal Dominant Inheritance (L1 x L1 mating)

pathological conditions which can be traced to a recessive allele. In order for such traits to appear in the phenotype, the genotype must be homozygous for the recessive allele since the presence of a single dominant allele will mask the affects of the recessive allele. Heterozygotes, then, may be carrying a recessive allele which can be transmitted to offspring, but such parents will not themselves show the trait. This fact often means great tragedy to the normal, healthy parents of a child born with a lethal genetic disease. One such condition, Tay-Sachs disease, or infantile amaurotic idiocy, involves progressive degeneration of cerebral function and results in the death of the affected child within a few years of birth. Until recently, prospective heterozygote (*Tt*) parents could not determine in advance of the birth of an affected child that they were, indeed, carriers of the allele which in homozygous (*tt*) combination was responsible for the condition. Now, techniques for distinguishing homozygous normals from heterozygous carriers should alleviate this problem. For most recessive traits, however, no diagnostic testing procedures are available.

The principles of autosomal recessive inheritance can readily be inferred from the information which has already been presented. All the children of two affected parents will also exhibit the condition. This follows from the obvious fact that affected parents must themselves have the homozygous recessive genotype combination and must, therefore, form gametes all of which carry the recessive allele. Parents of affected offspring, however, need not express the trait. Normal appearing parents of children with so-called "bird-head dwarf" syndrome (in which the child, male or female, has a small head, large eyes, beak-like nose, receding lower jaw and narrow face) are both heterozygotes (*Bb* × *Bb*), while the affected child is homozygous (*bb*) at this locus. As shown in Figure 2–5, unaffected children might also be born to such a couple, since the couple

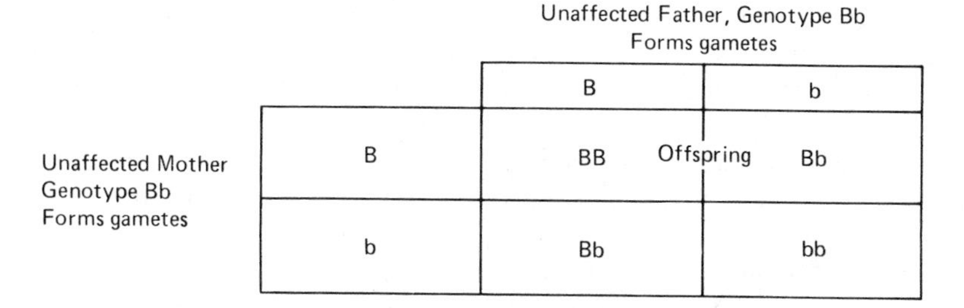

FIGURE 2–5 Inheritance of Bird-Head Dwarf Syndrome

could produce children with any one of three genotype combinations: *BB* (unaffected); *Bb* (unaffected); or *bb* (bird-head dwarfs).

Although the heterozygote is indistinguishable from the homozygote in many genetic traits, occasionally the heterozygote exhibits some of the properties of both homozygotes. Inheritance patterns of this type are described as involving *codominance.* Most of us are homozygous for normal adult hemoglobin, $Hb^A Hb^A$, but some people suffer from a condition called sicklemia, a form of anemia, which is due to homozygosity for the sickle hemoglobin allele ($Hb^S Hb^S$). The heterozygote ($Hb^A Hb^S$), however, forms both normal and sickle-cell hemoglobin and has a condition called sickle-cell trait which is usually harmless. In conditions involving codominant inheritance, both alleles have a phenotypic effect and the heterozygote can be distinguished from either of the homozygotes.

In addition to the genetic information carried on the twenty-two pairs of autosomal chromosomes, the sex chromosomes (*X* and *Y*) also carry genetic materials. These chromosomes not only play an important role in sex determination, as described in the previous chapter, but also carry genetic loci responsible for a number of phenotypic traits. The *Y* chromosome is now known to carry genetic instructions for the group of histocompatibility (*HY*) antigens, and perhaps for a few other traits, but many more genetic loci have been found to be located on the non-homologous portion of the *X* chromosome.

Sex-linked (more specifically, *X*-linked) inheritance follows a seemingly complex mode of genetic transmission, but the pattern is really quite simple so long as the behavior of the *X* and *Y* chromosomes in meiosis is kept in mind. You will recall that males transmit the *Y* chromosome only to males and the *X* chromosome only to their daughters, but mothers pass on an *X* chromosome to both male and female offspring. Some studies indicate that an inability to smell cyanide is due to an *X*-linked recessive allele. Since this trait is rare, affected males are likely to marry unaffected females. The male, who has only one *X*-chromosome, is neither homozygous nor heterozygous for this allele, but is referred to as being *hemizygous.* A hemizygous affected male would have the genotype $X^c Y$, but an unaffected female could be either homozygous ($X^C X^C$) or heterozygous ($X^C X^c$). As Figure 2-6 demonstrates, a hemizygous affected male married to a homozygous unaffected female could produce only normal children, boys or girls, but all daughters born to affected males will carry the allele, received from the father on the X chromosome he transmits to all his female offspring. A male with this condition, married to a heterozygous unaffected female, however, could produce both affected and unaffected daughters and affected and unaffected sons (see Figure 2-7). Affected daughters must receive the recessive allele from both parents, but affected sons receive the allele from the mother alone.

Certain traits, the expression of which is influenced by the sex of the individual, are carried on the autosomal chromosomes. One of these *sex-influenced* (rather than sex-linked) traits is pattern baldness. This condition involves the gradual thinning and recession of hair at the crown of the head and on both sides of the forehead. The most economical genetic explanation involves a single autosomal locus at which combinations of

	Affected Father (genotype X^cY)	
Unaffected Mother (genotype $X^C X^C$)	X^c	Y
	Offspring	
X^C	$X^C X^C$ (female)	$X^C Y$ (male)
X^C	$X^C X^c$ (female)	$X^C Y$ (male)

FIGURE 2-6 X-Linked Inheritance: Inability to smell cyanide (Father X^cY x Mother $X^C X^C$)

	Father	
Heterozygous Mother	X^c	Y
	Offspring	
X^C	$X^C X^c$ (female)	$X^C Y$ (male)
X^c	$X^c X^c$ (female) (affected)	$X^c Y$ (male) (affected)

FIGURE 2-7 X-Linked Inheritance: Inability to smell cyanide (Father X^cY x Mother $X^C X^C$)

two alleles direct the expression of the trait, depending on the relative levels of male and female sex hormones. Males with the genotype B_1B_1 do not show evidence of pattern balding, but those whose genotype is either B_1B_2 or B_2B_2 can expect an early demise of their hirsute glory. In females, however, pattern balding will not appear if the genotype is either B_1B_1 or B_1B_2. Only those women who are homozygous for the B_2 allele (B_2B_2) can be expected to develop the condition. The full explanation of this sex differential in expression involves the interaction of the genotype with the internal hormonal environment. In the presence of a relatively high ratio of male to female hormones, the phenotype for pattern balding will be expressed in both the homozygote B_2B_2 and in the heterozygote B_1B_2; where the female hormones are relatively dominant, the heterozygote genotype will not be expressed.

All of the traits discussed so far have involved "either-or" conditions —either you have Tay-Sachs disease or you don't; either you are a bird-head dwarf, or you're not. But many traits involve gradations of differences in expression. Skin color may range from dark brown to red brown to light brown, and right down to faintly pink; height may be 180 cm., 182 cm., 184 cm., and so on. The inheritance of these varying traits is more complex than those involving a simple single-locus system of inheritance. Many of these varying traits are the result of the expression of *polygenes*, or *additive genes*. A number of genetic loci, perhaps five or six, are thought to influence skin color, and the different combinations of alleles possible at each locus are thought to contribute, additively, small increments of the pigmentation of the individual. We can limit our considerations to a hypothetical model of two loci, *A* and *B*, at which combinations of the A_1 and A_2 and of the B_1 and B_2 alleles are present. Let us also stipulate that both the A_1 and B_1 alleles contribute two units to a scale of skin color, while the A_2 and B_2 alleles each contribute only one unit to the skin color continuum. The lightest skin color phenotype would be produced by the $A_2A_2B_2B_2$ genotype (color scale rating = 4) and the darkest phenotype by the $A_1A_1B_1B_1$ genotype (rating = 8). The marriage of two such double homozygotes would produce children, all of whom should approximate an intermediate scale reading of 6 (genotype $A_1A_2B_1B_2$). But, if two individuals with the $A_1A_2B_1B_2$ genotype married, they could each produce four kinds of gametes—A_1B_1, A_1B_2, A_2B_1 and A_2B_2; the offspring which could be produced are indicated in Figure 2-8.

The progeny from many such matings could be expected to fall into five color-scale classes: 8, 7, 6, 5, and 4. If a larger number of polygenic loci were involved, the distribution of the number of individuals belonging to the larger number of skin-color categories would approximate a normal curve distribution (Figure 2-9).

Gametes	A_1B_1	A_1B_2	A_2B_1	A_2B_2
	Offspring			
A_1B_1	$A_1A_1B_1B_1$ 8	$A_1A_1B_1B_2$ 7	$A_1A_2B_1B_1$ 7	$A_1A_2B_1B_2$ 6
A_1B_2	$A_1A_1B_1B_2$ 7	$A_1A_1B_2B_2$ 6	$A_1A_2B_1B_2$ 6	$A_1A_2B_2B_2$ 5
A_2B_1	$A_1A_2B_1B_1$ 7	$A_1A_2B_1B_2$ 6	$A_2A_2B_1B_1$ 6	$A_2A_2B_1B_2$ 5
A_2B_2	$A_1A_2B_1B_2$ 6	$A_1A_2B_2B_2$ 5	$A_2A_2B_1B_2$ 5	$A_2A_2B_2B_2$ 4

$A_1 = 2$
$A_2 = 1$
$B_1 = 2$
$B_2 = 1$

FIGURE 2-8 Polygenic Inheritance, Two-Loci, Two-Alleles Systems

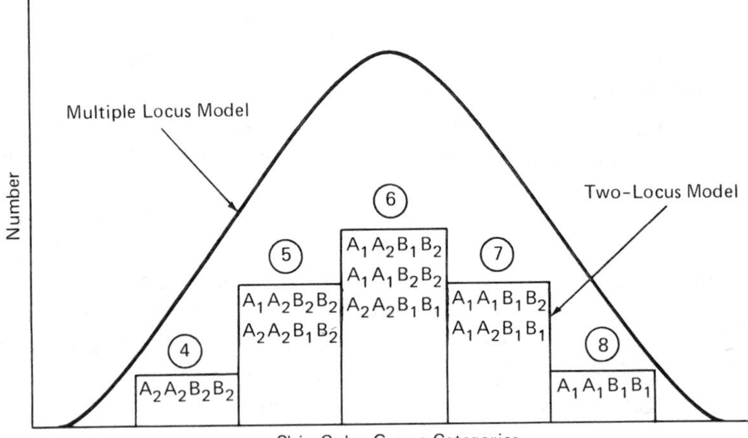

FIGURE 2-9 Distribution of Phenotypes in Hypothetical Polygenic Models

The study of many polygenic traits is complicated by the relatively great extent to which environmental influences affect their phenotypic expression. Constant exposure to sunlight can darken skin color; a high calorie diet can increase weight; disease and malnutrition can suppress growth and result in limited adult stature. But the modifying effects of environmental influence on the phenotypic expression of genetic traits is by no means confined to these polygenic traits. Genic action does not occur in a vacuum. The action of any gene involves the effects of other genes, beginning from the instant of fertilization (inter-genic actions) and extending to include the effects of the maternal environment (fetal-maternal interactions) and a whole range of internal (bodily) and external (extra-somatic) influences which we experience throughout the course of our lives.

One way of estimating the degree of environmental influence on the variation in expression of various genetic traits is to study the differences between genetic relatives reared in similar environments and those reared in contrasting conditions. Ideally, this could best be done by comparing *monozygotic twins* reared together and those in which members of each twin pair were separated at birth and raised thereafter in distinctive settings. Members of such twin pairs are genetically identical, formed from a single zygote which divides at an early stage of development to form two distinct, but genetically identical, embryos. If both members of these twin pairs showed little or no variance (concordance) in phenotypic expression despite differences in environment, then the trait is less subject to modification by environmental influences than are traits which do show differences (discordance) under these conditions. Unfortunately, the number of monozygotic twin pairs separated at birth and reared in differing settings is relatively limited, so we also examine *dizygotic twins,* formed as a result of the separate fertilization of two eggs, and *siblings,* who can also be expected to have one-half of their genes in common; we compare the variance in phenotypic expression of traits in these individuals when reared in the same household and when raised apart. It should be noted, of course, that siblings differ also in age, so that the younger sibling is raised in a home environment over a different time period and with an elder sibling who has already been established in the family. A brief selection from the results of many studies of this kind is presented in Table 2-1; the results indicate that concordance, implying a high degree of heritability, or genetic determination, exists for height and head dimensions, whereas weight, not unexpectably, is more subject to environmental modification.

Among the more striking forms of variation in phenotypic expressions of identical genotypes are those involving *incomplete penetrance* of a gene. The trait may appear in one twin but be lacking in the other member of a monozygotic twin pair; an individual may not express a trait although it is found in at least one of the parents and in the children of an unaffected individual. Very rarely, this might be explained as the product of a somatic mutation, but the incidence of these cases is far too high to justify explanation in terms of mutation alone. Rather, we should look to some event in the complex sequence of embryological development which results in the presence of the trait in most individuals (penetrance) but, in some cases, results in its absence in others carrying the appropriate genotype (non-penetrance).

incomplete penetrance – twin studies

TABLE 2-1 Average Difference Between the Two Members of Monozygotic Twin Pairs, Dizygotic Twin Pairs and Siblings, Reared Together and Monozygotic Twins Reared Apart.

Difference in	*Monozygotic Twins*	*Dizygotic Twins*	*Sibs*	*Monozygotic Twins Reared Apart*
Height (cm)	1.7	4.4	4.5	1.8
Weight (kilogram)	1.9	4.5	4.7	4.5
Head length (mm)	2.9	6.2	—	2.2
Head width (mm)	2.8	4.2	—	2.9

(From *Principles of Human Genetics,* Third Edition, by Curt Stern. W.H. Freeman and Company. Copyright © 1973.)

One such condition involves the appearance of small sinuses, cysts, or pits at the point where the upper ear attaches to the skin of the face. The condition is thought to be caused by the action of a dominant allele which prevents the complete obliteration of the branchial clefts. These clefts appear between the five branchial arches present in the human embryo from about the fourth week of development, but are usually obliterated by the sixth week of prenatal life in individuals with the homozygous recessive genotype. The trait (variously known as pre-auricular cysts, branchial cleft anomaly, or, more familiarly among anthropologists, as pili) occurs in about 5 to 10 percent of Australian aborigines, 5 to 7 percent in some American Indian tribes, and about 2 or 3 percent of Europeans and Americans.

A pedigree showing the transmission of the trait in a single kindred is shown in Figure 2-10, with affected persons shown by dark shading; the right or left half of the shaded symbol indicates which side of the head is involved. It should be noted that the male (II-2) does not express the trait, but it is present in his offspring. Accordingly, he must carry the responsible dominant allele and transmit it to his offspring, but the genotype has not "penetrated" his phenotype.

This pedigree also demonstrates another kind of variation in gene expression—*variable expressivity.* Of the affected individuals shown (I-2, II-4, II-5, and III-1), only two (I-2 and II-5) have the condition present on the right side. The other affected person in the second generation, II-4, exhibits the trait on both sides, while the single member of the third generation, III-1, has only the left ear affected. Thus, while the gene has penetrated the phenotype in all four affected individuals, it has shown variable expressivity in the degree of its manifestation when it is exhibited at all in the phenotype.

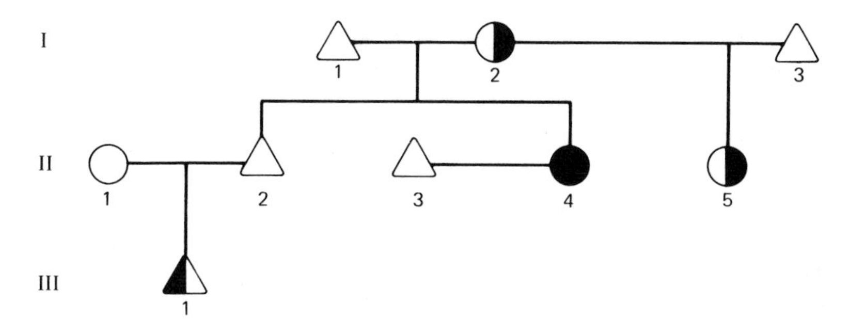

FIGURE 2-10 Pedigree Showing Transmission of Pili

Another approach to the study of variation in genetic expression comes from our understanding of the chemical nature of the genetic material and of how this material codes for the production of inherited traits. Clearly, there are many more known genetic traits in humans than there are chromosomes, so each chromosome must be a "package" of a number of genetic instructions. Were we to analyze a chromosome, it would be found to include a number of proteins (histones and protamines) as well as a double helical chain of a material called *deoxyribonucleic acid,* or *DNA*. In a series of experiments with lower organisms, investigators determined that it is the latter material, DNA, which transmits the genetic information necessary for directing cellular activities from one generation to the next. Each DNA molecule is formed from smaller units called nucleotides (see Figure 2-11), consisting of three components: phosphate; a 5-carbon sugar; and a nitrogeneous base. Any one of four kinds of nitrogeneous bases may be present: adenine, guanine, cytosine, or thymine. Two or more nucleotides are linked by weak hydrogen bonds between each pair of bases, while the sugar and phosphate groups form the "backbone" of a chain of nucleotides. Two strands of nucleotides, linked between bases, and wound around each other, form the characteristic double helical configuration of DNA.

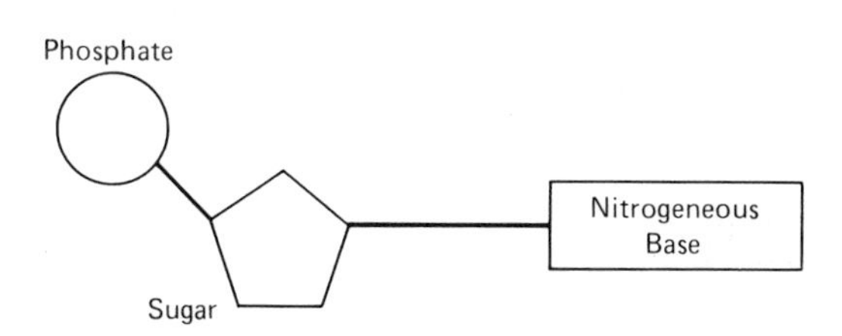

FIGURE 2-11 DNA Nucleotide

The four nucleotide bases of DNA do not pair at random, but only in the specific combination of adenine (*A*) with thymine (*T*), or guanine (*G*) with cytosine (*C*). This specificity normally ensures the exact replication of new DNA, for when it is formed, the two halves of the molecule separate and new sequences of bases attached to sugars and phosphates connect to the parental chain in the complementary order. Thus, if a four-base sequence of one half of a helical chain is represented by the base sequence *AGCT*, the other half of the chain must be *TCGA*. When the chain separates and begins replication, the half with the *AGCT* sequence will be matched by a newly forming complementary *TCGA* sequence, the parental *TCGA* segment with a new complementary *AGCT* chain (see Figure 2-12).

DNA has two functions—replication and the translation of genetic information into the directions for synthesizing the materials required by the cell. The latter function involves the production of *polypeptides*, chains of amino acids, which form *proteins*. Our bodies and the physiological activities which they carry out are dependent on proteins. Blood, for example, consists of about 99 percent protein substances. The chemical reactions required to sustain the living organism could not take place or could not proceed at a sufficiently rapid pace to maintain life in the absence of *enzymes*, special proteins which affect the rate of chemical reactions in living organisms.

The structure of DNA is as critical in the formation of proteins as it is in ensuring its own accurate replication. Proteins are composed of various combinations of amino acids, yet there are only some twenty amino acids; proteins composed of hundreds of amino acids may differ only by a single amino acid. Hemoglobin, a conjugated protein, consists of about 300 amino acids, yet normal adult hemoglobin differs from sickle hemo-

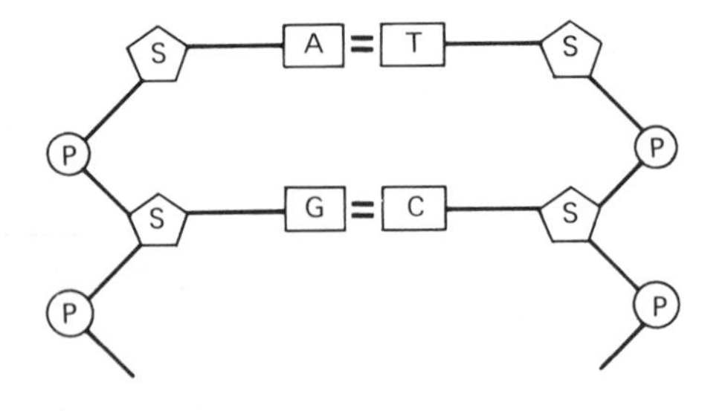

FIGURE 2-12 Nucleotide Chain, Unwound

S = 5 carbon sugar
P = phosphate
A = adenine
T = thymine
G = guanine
C = cystosine

globin only in respect to a single amino acid substitution. DNA functions in protein synthesis partly through its role in providing a template for the synthesis of another nucleic acid, *messenger ribonucleic acid* (mRNA), which then moves out of the nucleus to a structure, called the ribosome, where protein synthesis takes place. RNA is a single-strand molecule formed of (1) a phosphate, (2) a 5-carbon sugar, ribose, and (3) a nitrogeneous base. Three of the bases present in RNA (adenine, guanine, and cytosine) are identical with the bases found in DNA; RNA contains a fourth base, uracil, which replaces the thymine of DNA.

Transcription of the genetic message contained in DNA takes place after the DNA strands separate from one another. Through the mechanism of complementary base pairing, mRNA nucleotides are formed which follow the base sequence ordering specified by the DNA. Thus, if the DNA base sequence is *ACC*, the complementary mRNA sequence will read *UGG* (remembering that uracil in RNA substitutes for the thymine of DNA). A chain of mRNA, complementary to the sequence of nucleotides in DNA, is formed and the mRNA chain detaches from this DNA template and leaves the nucleus; see Figure 2-13.

As the messenger RNA passes to the ribosome, outside the nucleus of the cell, another nucleic acid, *transfer RNA*, transports amino acids to the ribosomal site of protein synthesis. One end of the tRNA molecule attaches to a specific amino acid, while the other end connects only to that three-letter sequence of messenger RNA specific to the particular

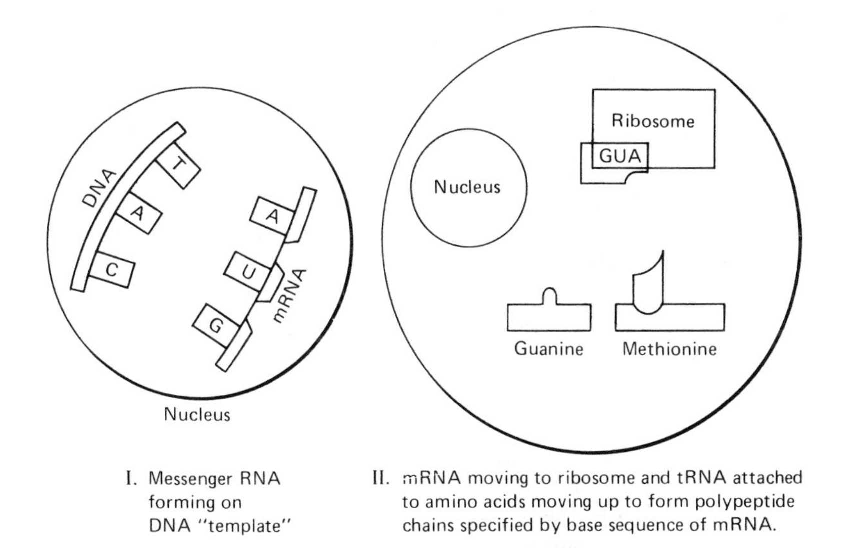

I. Messenger RNA forming on DNA "template"

II. mRNA moving to ribosome and tRNA attached to amino acids moving up to form polypeptide chains specified by base sequence of mRNA.

FIGURE 2-13 Schematic Diagram of Protein Synthesis

amino acid. It has been determined, for example, that the three-base sequence ("triplet codon") of DNA, which reads *TAC*, specifies the formation of the sequence *AUG* by the messenger RNA. At the ribosome, the matching tRNA will transport the amino acid methionine, and only methionine, to the segment of the mRNA chain which has the sequence *AUG*. As a chain of mRNA passes to the ribosome, the sequence of amino acids specified by the mRNA sequences is brought together by the tRNA molecules, and peptide bonds are formed between adjacent amino acids to produce a polypeptide chain.

Many inherited traits are the product of the expression of genes which code for a specific protein or enzyme, but some characteristics entail the sequential operation of a number of loci. Prime among these conditions are the various "inborn errors of metabolism" first described by Garrod. Melanin pigment is produced from the breakdown of tyrosine which is acquired, in part, from the breakdown of ingested proteins containing the substance, and in part from the metabolism of another amino acid, phenylalanine. In the normal pattern, phenylalanine is metabolized in the presence of the enzyme phenylalanine hydroxylase to form tyrosine, and tyrosine is further metabolized with the aid of the enzyme tyrosinase into a precursor of melanin. If tyrosinase is not formed, melanin is absent, and the individual exhibits the characteristics of one form of albinism. If the metabolic block occurs earlier, at the point of phenylalanine metabolism, the affected victim suffers from a rare hereditary disease, phenylketonuria. Since tyrosinase may be present in the victims of this disease, some melanin production from ingested tyrosine takes place, but it is limited and the victims are characteristically light in pigmentation. In addition, those affected with phenylketonuria accumulate high levels of phenylpyruvic acid and this substance blocks the conversion of the amino acid tryptophan to serotonin, a substance in the brain essential to normal mental functioning. Fortunately, early diagnosis can result in the amelioration of these symptoms if the victim of the disorder is placed on a low phenylalanine diet from birth (see Figure 2-14).

As Figure 2-14 shows, a metabolic block anywhere along the sequence of biochemical reactions leading to the production of melanin can affect the pigmentation of the affected individual. In some cases, blocks at different steps in the synthesis of products can produce similar results (*gene mimics*). Several forms of albinism are known; one form involves recessive inheritance, one involves dominant inheritance, and at least two forms of albinism are differentiated on the basis of the presence or absence of the enzyme tyrosinase.

Some genes are also said to have *pleiotropic effects;* that is, a single

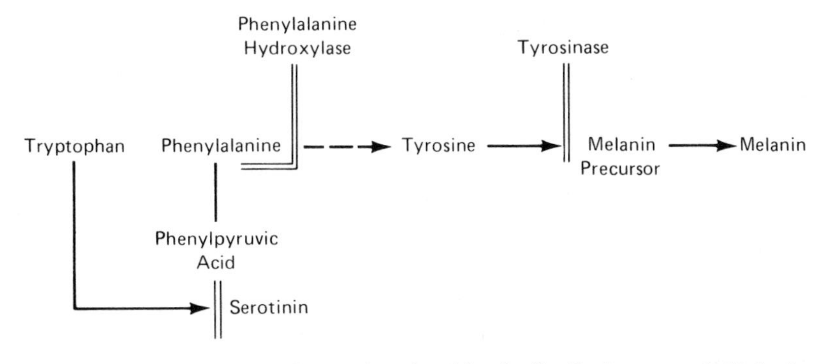

FIGURE 2-14 Some Steps in the Metabolic Pathways of Melanin Production

locus may affect the expression of several characteristics. In phenylketonuria, light pigmentation, characteristic odor, and mental retardation all result from the action of a single allele when present in the homozygous recessive condition. In one form of arachnodactyly, Marfan syndrome, the presence of a dominant allele entails defects in the lens of the eye, excessively lengthy fingers and toes, and characteristic defects of the heart.

In contrast, some traits involve the interaction of several genetic loci (*epistasis*). Persons who are *O* blood type have an *H* substance which is formed from the transformation of a Le^a substance (coded for by the gene for the Lewis blood group antigen) by an enzyme that adds a sugar to the Le^a substance molecule. However, the Le^a antigen is present on the red blood cells only when the genotype is also homozygous for the recessive allele, *se,* of the secretor system which affects the presence or absence of the *ABO* blood group antigens in the saliva and other bodily secretions.

The complexities of inheritance and genic expression are part and parcel of our growing understanding and appreciation of the rich fabric of human diversity. The anthropologist cannot be a full time geneticist, for the field of genetics requires its own commitment, dedication, and expertise. But the anthropologist needs a fundamental knowledge of the genetic foundations of human variability in order to identify the kinds and effects of biocultural interactions which, uniquely, affect human evolution. The basic principles of Mendelian genetics outlined in this chapter are that necessary substructure for our further study of human variation, not only in the individual, but also in the societies of people who inhabit our world.

MENDELIAN GENETICS—PROBLEMS

1. A woman is wooly haired, with the genotype Ww. Her husband is homozygous recessive at this locus and has normal hair. What is the chance that wooly hair will appear in:
   (a) her first child?
   (b) her first and second child?
   (c) all her six children?
2. If a victim of phenylketonuria marries a normally pigmented woman, what are the possible phenotypes and genotypes of their children?
3. A man with "classic hemophilia," a sex-linked trait, married a normal woman whose father was also hemophiliac. Could this couple have a daughter who suffers from this disease?
4. A man who is heterozygous for Nordic eyefold (Nn) and pattern baldness (B_1B_2) marries a woman who has normal eyefolds (nn) and pattern baldness. What is the probability that their first child will be a boy with normal eyefolds and pattern baldness?
5. A heterozygote man (Jj) who entertains his children with his jaw winking almost died from a G6PD reaction to the moth balls in his fiance's closet (G6PD deficiency is an *X*-linked trait). After his recovery he married her, even though she couldn't entertain her new step-children with jaw winking and despite the fact that her family was very big in dry cleaning. Given the fact that no one in her family ever had a G6PD reaction to her closet, can this couple have a son with G6PD deficiency who winks his jaw?

REFERENCES AND RECOMMENDED READINGS

HARTL, D.L. 1977. *Our Uncertain Heritage: Genetics and Human Diversity*. Philadelphia: J.B. Lippincott Co.

LEVINE, R.P. 1968. *Genetics*. 2nd ed. New York: Holt, Rinehart and Winston, Inc.

McKUSICK, V.A. 1969. *Human Genetics*. 2nd ed. Englewood Cliffs: Prentice-Hall, Inc.

McKUSICK, V.A. 1975. *Mendelian Inheritance in Man. Catalogs of Autosomal Dominant, Autosomal Recessive and X-linked Phenotypes*. 4th ed. Baltimore: Johns Hopkins University Press.

STERN, C. 1973. *Principles of Human Genetics*. San Francisco: W.H. Freeman and Co.

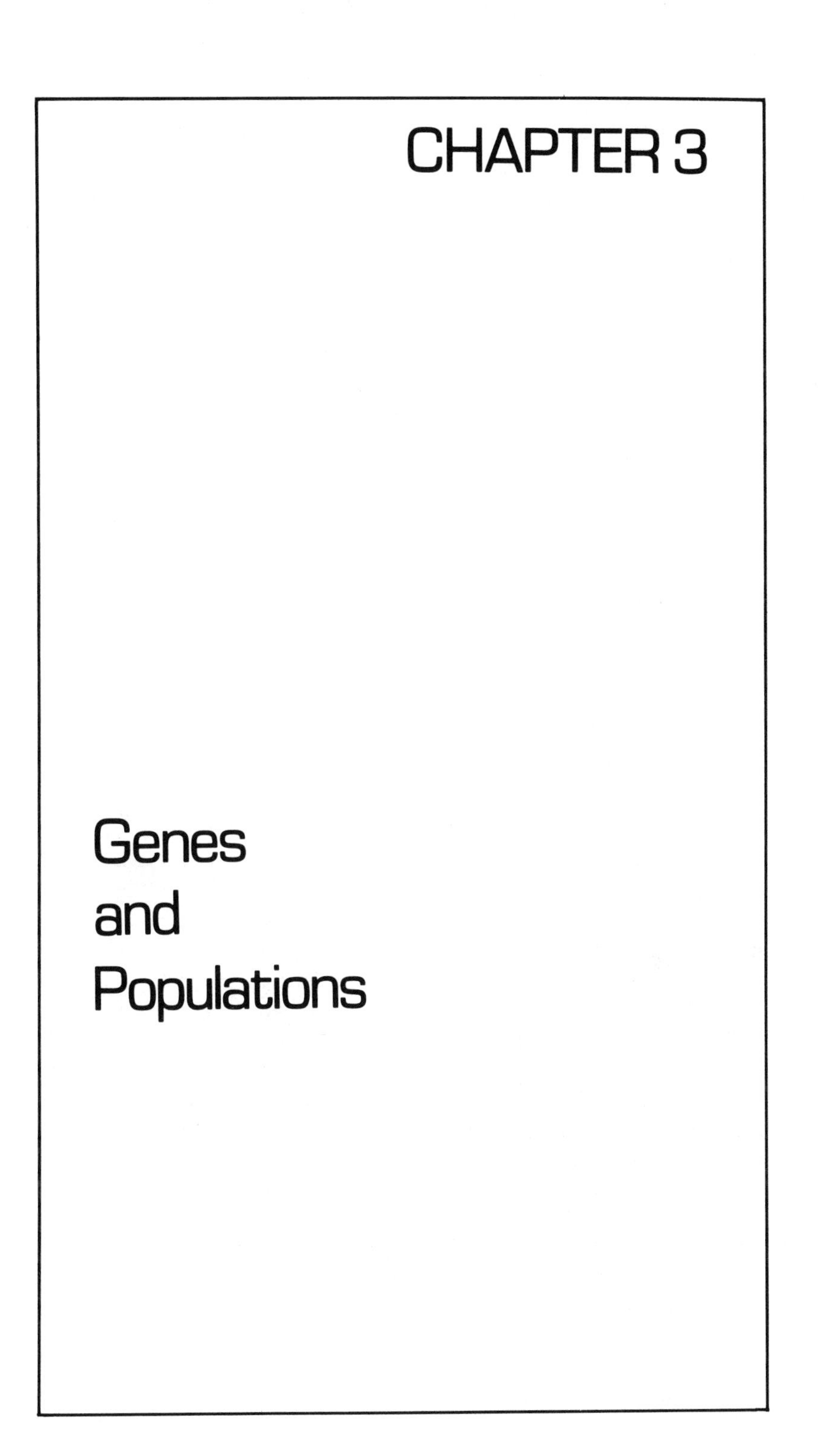

CHAPTER 3

Genes and Populations

Individuals may change, but they do not evolve. Evolution, as defined in modern theory, refers to non-cyclical changes in the allele frequencies of a population's gene pool from one generation to another. While a population is composed of individuals, the properties of the population are more than the sum of the individual members or their activities. You don't have a gene pool, but you can contribute to the gene pool of the population to which you belong. In much the same way, we can just as well say that an individual is more than the sum of his or her constituent parts. I am not just two arms, two legs, a body, and a head, but the (usually) coordinated whole system resulting from the interactions and integrations of various component parts. The properties of that system, as of any integrated system, cannot be comprehended solely in terms of the constituent parts of that system. But, before we try to pursue the study of the population as the fundamental unit of evolution, it is important to understand the meaning and applications of the term population itself.

Although the word has broader applications, our interests are served by limiting our concerns to the *biological population,* a group of interacting individuals belonging to a single species. A group of these *conspecifics,* or members of a single species, occupying a delimited area may constitute a census population. By counting these individuals and observing their distribution, recording the numbers of births and deaths which occur over a given time period, and noting the numbers of individuals who enter or leave the group, we can describe certain attributes of the population: birth rates, death rates, migration rates. Note that an individual does not have a death rate, since he or she only dies once. These vital rates are the attributes of the population.

A group of conspecifics among whom mating occurs, or, at least occurs more often than with members of other such groups, constitutes a *Mendelian,* or breeding, population. This group has a *gene pool,* the genetic materials temporarily distributed among the members of the breeding population as a set of genotypes. Again, it should be clear that an individual has a body of genetic material, the *genome,* from which rare contributions are made to the ongoing gene pool of the population, but the individual's genetic materials have no evolutionary value until and unless these are incorporated into the gene pool of succeeding generations.

A species may now be seen as composed of one or more Mendelian populations, and would include all those individuals and groups among whom mating may actually or potentially occur. Although occasional exceptions are known, a biological species is reproductively isolated from all other species. In a widely distributed species, such as humans, constituent populations are subject to a wide range of environmental differences and stresses and have experienced vastly different histories. Breeding between a native of Greenland and an inhabitant of sub-Saharan Africa is biologically feasible, but unlikely to happen because of the distance separating the two groups. Thus, groups of populations with more closely related ancestry and which have evolved in a common region are likely to share genetic and morphological similarities to a greater degree than they would with populations which are more dispersed and with which they have had less opportunity for genetic exchange. Populations with a greater degree of shared heritage and environmental contiguity are sometimes grouped together and treated as a unit, a subdivision of the species, and have been variously termed a variety, a race, or a subspecies.

But since evolutionary pressures and genetic changes are expressed at the level of each population's gene pool, the basic level of research focuses on the population rather than on the species as a whole or any of its larger subdivisions. This is particularly true in the case of humans, since even the mating systems of different social groups vary greatly and have such potential impact on the microevolution of human populations. So far as we know, no population of fruit flies practices preferred patrilateral cross-cousin matings (in humans, the mating of a man with his father's sister's daughter), but this occurs in some human populations and has genetic consequences, particularly as to the transmission of *X*-linked traits. In studies of human populations, we can often make use of the concept of the *deme,* or localized breeding population, but even here it is necessary to approach with caution. Insofar as mating occurs primarily among its members, the international jet set is a Mendelian population, but it can scarcely be described as localized in space. On the other hand, a localized population need not constitute a breeding population: consider the residents of a monastery as an extreme example, or the numerous endogamous, or in-marrying, ethnic groups of a large city as a more plausible objection.

For all practical purposes, the population, as we will use the term here, can be considered as the (localized) breeding group and can be seen to have certain attributes—a gene pool and population dynamics—which are not merely the sum of its members or of individual events, but are a property of this hierarchical level of organization. The study of these attributes comprises the content of the remainder of this book; we can begin with an introduction of the relationship between individuals and the events experienced by individuals to the gene pool and the dynamics of the population.

From previous chapters, it should now be clear that many observed differences between populations or among individuals within a population have a genetic basis. Thus, the phenotypic diversity found in a population expresses, but only in part, the genetic variation characteristic of the group. Each of us has a genotype and expresses only a part of the potential genetic variability in the phenotype. I am a heterozygote for one of the loci involved in the *Rh*-negative blood type, but my phenotype is that of *Rh*-positive blood type. However, I can transmit to my offspring either the allele for *Rh*-positive or the allele for *Rh*-negative blood type, so my child's phenotype depends also on the allele received from the other parent. Therefore, the genetic constitution of the next generation depends on the genetic potential of the whole gene pool of my generation. This genetic material is temporarily dispersed in the genotypes of the individual members of the group, but the gene pool of the population will consist of all the potential genetic contributions which can be made by its members.

Perhaps this can better be appreciated by considering how this might operate in practice. Suppose we select a group of 200 adults to establish a pioneer space colony, and we examine the *MN* blood locus in our attempt to understand the relationship between individual genotype and the gene pool of a population. Among the criteria for selecting the members of this group, we would probably want to specify that a range of variation is represented, so we stipulate that the crew must contain fifty people with the *M* phenotype, 100 with *MN* phenotype and fifty with *N* blood type. Since *MN* is a codominant trait, we can even specify the genotype from the phenotype of each person: individuals with *M* blood type have the genotype *MM;* those with *MN* phenotype are *MN* genotype; and *N* phenotypes have the genotype *NN*. We also ensure sexual equality in this respect by ensuring that half of each phenotype group must be male and the other half female. As consultants on this project, we prepare a roster for the scientific officer of the crew, and it would look something like Table 3-1.

These individuals form a population which includes the genetic variation represented by the genotypes of all its members: 25 percent of all individuals carry only the *M* allele; 25 percent only the *N* allele; but 50 percent of all individuals carry both the *M* and the *N* allele; so, half the alleles represented in these genotypes are the *M* allele, while the other half are the *N* allele. The gene pool of the population consists of 50 percent *M* alleles and 50 percent *N* alleles; or, as it is usually expressed, the gene pool for this locus includes *M* allele with a frequency of 0.50 and *N* allele with a frequency of 0.50.

This intangible gene pool, the totality of genetic potential at all loci, transported in the genotypes of these pioneers, attains continuity by transmission to an offspring generation. If that transmission is not precise, then evolution—changes in the allele frequencies of the gene pool—has taken place. Such changes may be due to one or more of several evolutionary forces, of which the more important are mutation, gene flow, genetic drift, and natural selection.

Mutations, any changes in the genetic material (except gene recombinations produced by chromosomal crossing over), are of evolutionary

TABLE 3-1 Crew Roster: Genetic Log

Phenotype: Blood Type:	*M*	*MN*	*N*	
	MM	*MN*	*N*	*Totals*
Number:				
Males	25	50	25	100
Females	25	50	25	100
Totals	50	100	50	200
Frequencies				
Totals	.25	.50	.25	1.00
Gene Pool Frequencies	M = .50		N = .50	

significance only when these affect germinal cells. If our hypothetical colonists should be exposed to ionizing radiation or chemical mutagens which cause extensive bodily change, such alterations would be of interest, but insignificant from an evolutionary perspective unless germinal cells are altered. Any mutations affecting the gametes could be transmitted to the next generation, and this would again mean a change in the allele frequencies of the gene pool.

If our space travelers were to locate a colony of humans on some other planet (a not implausible event in the future) and mate with members of the other group, one kind of gene flow, intermixture, would take place. Another kind of gene flow, the movement of some of our colony into some other environment, could also affect the characteristics of that group as the migrants are exposed to new living conditions.

One form of genetic drift, founder effect, has really been precluded from operation by our insistence on the representativeness of the colony members. When a small group of founders is derived from an ancestral population, these individuals may not exactly represent the entire range of genetic variation present in the larger parental group. Since the founders are relatively few in number, by chance alone we might have picked individuals all of whom were *M* blood type for our colony. In that case, the gene pool of the population would have had a frequency of 1.00 for the *M* allele and 0.0 frequency for the *N* allele, and subsequent generations would have lacked the *N* allele (or the *MN* or *NN* genotype) unless a rare mutation occurred or until genetic exchange with a genetically divergent or more diverse population took place. However, intergenerational gamete sampling–random differences in the contributions of some parents to the gene pool of the succeeding generation–could well be expected to occur in this small, isolated colony. Over enough generations, this might even lead to the loss of one or the other allele from the gene pool. We could well expect this latter form of genetic drift to play a role in the microevolution of our space colony's descendants.

Finally, natural selection could operate to change the allele frequencies of the gene pool. Again, looking only at the single *MN* locus, if individuals with any one of the three possible genotypes is advantaged by virtue of the possession of the respective trait in contributing offspring to the effective population of the next generation, that trait would have a selective advantage. For example, it has been suggested that heterozygous (*Tt*) carriers for Tay-Sachs disease may have had greater resistance to typhoid fever or tuberculosis in the historic past. If more of these heterozygotes survived to reproduce while all homozygous recessive individuals (*tt*) died in early childhood and some homozygotes (*TT*) were more susceptible to death from typhoid fever or from tuberculosis, the persistence of the carriers in unexpectedly higher frequencies could be explained. From findings in a number of populations, we might expect to find a larger number of *MN* offspring in the space-born generation than among the original colony. There are various ways by which selection operates—through fertility differentials, through survival differentials, through mortality differentials—but the essential process is one of nonrandom differential contributions of some genotypes to the gene pool of succeeding generations. If selection operates so as to alter the allele frequencies underlying the traits which prove more or less reproductively fit, the resultant changes constitute evolution through natural selection.

A second attribute of the biological population—the growth, or dynamics of the population—concerns those very demographic processes (fertility and mortality) which are intimately involved in the transmission of genetic information to succeeding generations. At a moment in time, the structure of a population can be described in terms of the age and sex composition of the group, and this is usually portrayed by a *population pyramid* (Figure 3-1). The description is static, but the figures are the result of birth, death, and migration rates which have already been experienced. As a result, descriptions of this kind can be misleading: *Cohorts,* persons all born during a certain time period, may number 1000 at time X *either* because 1000 individuals were born during this period and all survived to be counted at time X *or* because 2000 persons were born in this interval but only half of them survived long enough to be counted. At best, a population pyramid is a gross diagrammatic representation of the age-sex structure of a population at a given time, and is the outcome of a variety of combinations of demographic processes which have operated in the past.

In 1799 Thomas Malthus wrote *An Essay on the Principle of Population . . .*, which argued that human abilities to increase food supplies are less than our capacities to reproduce; unless human population growth is subject to preventive checks (moral restraint, delayed age at marriage, birth control) or positive checks (war, disease, infanticide), the potential geometric increase in the number of people must necessarily outstrip the essentially arithmetic potential increase in the means of subsistence. This statement is, of course, a greatly reduced summation of a lengthy argument which Malthus refined in several editions and additional essays, but these essential points can provide a useful model for examining the nature and history of human population growth.

Theoretically, a human female may release a viable ovum during every menstrual cycle throughout the course of her reproductive life span of perhaps thirty years, unless pregnancy occurs. If we allow about thirteen ovulations per year, this would amount to slightly less than 400

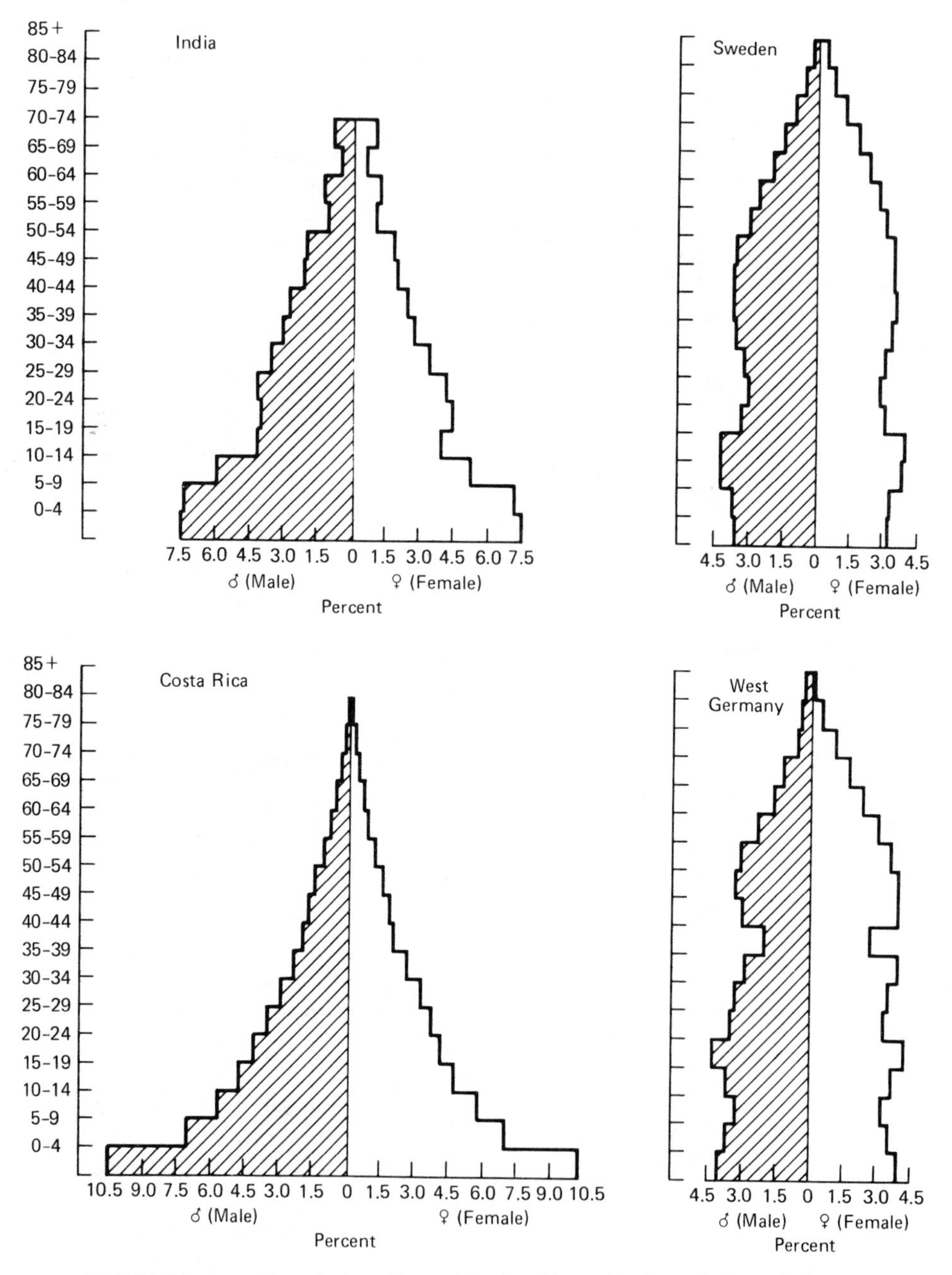

FIGURE 3-1. Population Pyramids for Four Nations (adapted from Bogue, 1969, p. 151).

potentially pregnable gametes formed. On the other hand, a human male may produce viable sperm throughout the course of his adult life; so we can focus on the delimited period of female *fecundity* (biological capacity to reproduce) as the more critical aspect of human *fertility* (actual reproductive performance). In fact, female fecundity is lower than the physiological maximum for a number of reasons. Some menstrual cycles may not involve the release of an ovum (anovulatory cycles), particularly during the early and late stages of the woman's reproductive life period. Ovulation is, of course, suppressed during pregnancy and in the period immediately after delivery, and to some extent while the mother is lactating, nursing the child. Hormonal disturbances, certain disease conditions, even emotional stress, as well as severe nutritional deficiency, may also affect menstrual cycles and ovulation. It follows that female fecundity never attains to the theoretical predicted maximum. If this alone is not sufficient cause to question extreme claims about human reproductive potential, consider the additional factors which can affect human fertility.

Davis and Blake (1956) have suggested that human fertility is affected by three kinds of factors. The first of these, the "intercourse variables" factor, comprises the conditions and practices which limit sexual opportunities. Although a woman may be fecund from the age of about twelve until she approaches menopause, considerable periods of her life may be spent without sexual activity. Age at marriage or intercourse may be delayed until a considerable part of the reproductive years has passed; there may be lengthy periods of abstinence in conformity with prevailing sexual taboos; remarriage may be delayed or not take place at all after the death of a former spouse or after divorce. Nag, who carried out studies of factors influencing fertility in sixty-one nonindustrial societies, found that the relaxation of sexual taboos in recent years has played an important role in the increased fertility levels observed in many of the groups he examined.

The second category, conception variables, includes the practices associated with birth control as well as other conditions and practices which may limit conceptions. Contraception is not a singular property of technologically advanced societies, but has been practiced throughout much of human history and in a wide range of human societies. The methods vary in technique and efficacy, but the results, intended or realized, are to limit reproductive performance. In addition to contraceptive practices, diseases, particularly venereal diseases, have made significant contributions to reductions in human fertility.

Finally, Davis and Blake draw our attention to variables associated with gestation and parturition as factors which can influence fertility. If contraception fails, abortion or early neonatal infanticide can reduce the fertility levels of a population—with the added factor of having lowered the fecundity of the female during the term of the pregnancy. In addition to these practices, fertility will be influenced by the level of medical care and by general living conditions which can affect the health and survival of the pregnant woman and the fetus during pregnancy and delivery.

With all these factors to be considered, it may seem surprising that women ever do have children. Even when ovulation occurs, the ovum is viable only for a brief period of perhaps forty-eight hours, and sperm must be available and capable of penetrating the rather impressive barriers around the egg during this limited time. Since frequent intercourse may lower the sperm count, and since infrequent intercourse results in the ejaculation of overmature sperm, timing of intercourse during this period may also be critical. It has even been estimated that as many as 50 percent of all conceptions in humans result in spontaneous abortion. From this viewpoint, the problem seems not to be one of unchecked, rampant fertility, but of the sheer unlikelihood that conception will occur and result in the delivery of a viable child.

But women do become pregnant and deliver healthy children who survive to produce their own children. The rate at which they do so has usually been below that which would lead to geometric rates of population growth, given the mortality rates which have prevailed for much of human history. Some idea of the maximal levels of reproduction in human populations may be obtained from studies among Hutterite women (Eaton and Mayer, 1954). The Hutterites, members of one of the small Anabaptist sects that came to America in the eighteenth and nineteenth centuries, place a high value on children and provide communal care for the children of each Hutterite colony. The population size of the Hutterites has grown at a prodigious rate since their arrival in the New World, particularly in this century—in part because of high fertility rates, but also because excellent medical care is provided without cost to the individual, and mortality rates have been correspondingly low. Eaton and Mayer found in their studies that the average number of children ever born to women who had completed their reproductive lifespans was slightly less than eleven. This figure probably represents a realistic estimate of the maximum fertility level achievable by females in human populations.

Fertility levels, as the Hutterite case suggests, are a function of prevailing morbidity and mortality conditions. Most Hutterite females at birth will survive to a respectable age, but even a large cohort of females at birth may fail to contribute a large number of offspring to a growing population if few survive to reach reproductive maturity or if they are subject to diseases which reduce fecundity. A comparison of major causes of death in the United States in 1900 and 1975 reveals that infectious diseases (respiratory infections, gastrointestinal diseases, diphtheria, etc.) removed many children from the population well before they could have made any contribution to the gene pool. Today, the major causes of death—cancer, cardiovascular diseases, cerebrovascular conditions—are more likely to affect those whose genetic contributions have already been made.

Although a doubling or tripling of the population each generation has occurred in a few societies and over relatively few generations, the balance between mortality and fertility schedules in most societies has limited population growth to lower rates. According to the Population Reference Bureau's *1975 World Population Data Sheet,* the estimate of the number of years required to double population at existing growth rates for the entire world's population is thirty-six years; this value varies from ten years for Kuwait to 347 years for Finland, Austria, Luxembourg, East Germany and Malta. Obviously, even in today's world most populations are not increasing at a geometric or exponential rate, and increases in food supply have far exceeded any advances envisaged by Malthus. There is some evidence that the trend to higher birth rates, which followed declining death rates in developing countries, is beginning to recede. This will not end population growth: under improved mortality conditions, more infants survive to reproduce themselves; but rapid growth rates in these countries during the past few decades will probably not continue at the same levels. The genetic consequences of these changes are to reduce the role of mortality differentials in selection and to place greater weight on variations in fertility.

The population principle of Malthus has often been misunderstood and maligned. For our purposes, it is best seen as a model which, because it is a simplified system, raises questions as to its non-correspondence with reality. Human populations do not grow at an exponential rate, at least not over long periods of time, because a number of biological and cultural variables, operating together and independently, limit human fertility below the theoretical maximum which geometric growth rates must assume. Malthus was aware of some of these restraints, and others have extended our understanding of their complex nature and operations. Genetic evolution occurs as a result of differential fertility, survivorship, migration, and mortality. These demographic processes, characteristic of Mendelian populations, are the cornerstone of human microevolution.

The model of genetic adaptation presented above provides a powerful key to the appreciation of human variation, but little attention has yet been given to the wide range of phenotypic expressions of specific genotypes which are available to humans as a means of responding to environmental stress. Human adaptability may entail direct allele frequency changes as a result of natural selection, but one of the long-term trends in human evolution has been the capacity to vary the pattern of behavior and functioning in response to stress. The culminating feature of this trend is represented by the development of culture. The Luiseño Indians of southern California have not undergone genetic selection for the enzymes required to digest the toxic chemicals present in acorns, but they have developed (cultural) techniques for removing these toxic elements from one of their dietary mainstays. But some of the mechanisms for adjusting to environmental stress are provided through developmental and physiological plasticity. This is not to say that such trends are not, themselves, genetically determined responses; indeed, the genotype determines the capacity of the individual to respond through such adjustments. However, such adjustments cannot readily be connected with specific genotypic differences, and it is useful to consider the whole range of response to stress as part of human adaptability—including specific genetic adaptation as well as adjustments through phenotypic plasticity.

An example of morphological and physiological adjustments in human populations subject to environmental stress is provided by the complex of responses found among groups living in high altitude regimes. At altitudes of 4000 meters and above, each inhalation includes only up to 60 percent as much oxygen as would be inhaled at sea level atmospheric pressure. Yet, large and numerous populations in the Himalayan mountains of Asia and in the South American Andean highlands live at such heights. Even the short-term resident of the high altitude environment will respond to oxygen deficiency with a variety of mechanisms that will lead to physiological homeostasis—by rapid breathing which shifts the blood toward an alkaline level, which in turn stimulates an adrenal response and leads to the release of stored blood cells into the blood stream. This short-term increase in hemoglobin (carried in the more numerous red blood cells now in circulation) along with changes in heart rate, ion excretion by the kidneys, lung capillary dilation, and other adjustments provides a first-line series of adjustment to hypoxic stress. Over a longer period of residence, these short-term responses are modified even further, but such changes are reversible upon return to lower altitudes.

Populations with a long history of residence at high altitudes exhibit pronounced differences, physiologically, from their lowland relatives, and some of these traits (hypertrophy of the right ventricle of the heart, thickening of the pulmonary blood vessels, increased anterior-posterior thoracic diameter) are not usually reversible. High altitude populations living in the most remote parts of the world commonly resemble each other in a number of these features, even in the absence of any evidence of close genetic relationship. These similarities suggest that some common genetic attributes underlying such traits are selectively advantageous in all populations living under this kind of environmental stress. Minor differences in the adjustment patterns of Himalayans and Andeans to high altitude stresses do exist, and differences in the responses of different populations to cold, heat, even in susceptibility to common infectious diseases, suggest that genetic differences among populations can also provide somewhat different or alternative responses to similar stress stimuli.

Organisms are the product of heredity-environment interactions, and while introductory genetics books must necessarily speak of *the* phenotype which results from a particular genotype, the fact is that no trait develops in isolation. Some traits, such as the blood group antigens, involve relatively few steps from genic action to phenotypic expression; the hereditary influence is relatively greater than in traits, such as stature and weight, which are subject to considerable environmental influence during development. It is possible to study phenotypic variability within a population and to estimate the ratio of genetically caused variation to the total variation of a character in a population. This procedure entails the partitioning of such variation into two components: that part of the variation due to genetic differences, and the part due to environmental differences. However, the results are specific to the variance in each population and cannot be extrapolated to other populations. Failure to recognize this fundamental fact has needlessly exacerbated many discussions of race and intelligence.

In summary, then, Mendelian populations are the fundamental units of evolution and have certain attributes, a gene pool and population dynamics, which provide, respectively, the level at which evolution works and the demographic mechanism of evolutionary change. The model of genetic adaptation envisions changes in allele frequencies in response to environmental pressures, primarily through the operation of natural selection on the gene pool of the population through nonrandom differential contributions of some genotypes to the gene pool of succeeding generations. New materials are added to the gene pool through mutation, and new genotype combinations may appear as a result of intermixture of formerly distinct populations. Random genetic drift, whether as a result of sampling non-representativeness of a founder group, or in consequence of intergenerational gametic sampling variance, is limited to small populations or isolates. These evolutionary processes operate through such demographic events as the births and deaths occurring in the population, and these determine whether and what kinds of genetic materials are transmitted to the effective population of succeeding generations. The whole range of genetic variation, from traits with a high degree of heritability to those whose phenotypic expression is subject to considerable environmental influence, is subject to these same evolutionary processes.

Population Genetics

CHAPTER 4

Introduction to Population Genetics and the Hardy-Weinberg Principle

Mendel's use of the term "dominant" to refer to those units of heredity whose effect was expressed in the phenotype whether present in like or unlike pairs (we would now say "whether present in homozygous or heterozygous genotype combination") had some interesting consequences. As students of language might have predicted, the connotations of the term were somewhat misleading; dictionary definitions usually used such synonyms as "prevailing," "commanding," and the like. Thus, it wasn't really surprising that the question arose as to whether dominant conditions would not prevail while recessive alleles, and the traits expressed by the homozygous recessive genotype, would inevitably become uncommon and eventually disappear. This issue was resolved in part when, in 1908, the mathematician Hardy and an obstetrician, Weinberg, separately formulated what came to be known as the Hardy-Weinberg model, which forms the foundation of modern population genetics.

You will recall from previous chapters that when two heterozygous (*Hh*) parents mate, they can form three kinds of zygotes: *HH, Hh,* and *hh.* This can be shown diagrammatically, as in Figure 4-1.

If half of the gametes produced by the father carry the *H* allele and the other half carry the *h* allele, and the same is true of the gametes produced by the mother, we can even predict the probability with which each

		Father (Hh Genotype) Forms Gametes	
		H	h
Mother (Hh Genotype) Forms Gametes	H	HH	Hh
	h	Hh	hh

(Zygotes:)

FIGURE 4–1

zygote combination is likely to occur (see Fig. 4–2). In summary, the probabilities would add to 1/4 *HH*, 1/2 *Hh*, and 1/4 *hh*. This relationship can be expressed by a simple formula:

$$(\text{Probability of } H + \text{Probability of } h)^2 = (1/2H + 1/2h)^2 = 1/4\ HH + 1/2\ Hh + 1/4\ hh.$$

Since we sometimes look at the *H* locus, at other times at other loci, we can generalize this calculation to any bi-allelic loci by using the letter p to refer to one allele and q to refer to the other allele at a locus. We can then express the general principle as:

$$(p + q)^2 = p^2 + 2pq + q^2 = 1.00.$$

Now, let us consider applying this model to a population in which the allele frequencies are $p = .80$ and $q = .20$. If mating is at random in respect to this trait, and no evolutionary forces are operating at this locus, then we can predict that the genotype distributions will be $HH = 0.64$, $Hh = 0.32$, and $hh = 0.04$, and these frequencies will sum to 1.00. Conversely, in a population of 100 with these genotype distributions, the allele frequencies can be calculated by counting:

64 people have two *H* alleles each = 128 *H* alleles
32 people have one *H* allele each = 32 *H* alleles

160 *H* alleles

4 people have two *h* alleles each = 8 *h* alleles
32 people have one *h* allele each = 32 *h* alleles

40 *h* alleles

Thus, the frequency of the two alleles is:

H = 160/200 (total alleles in the population) = 0.80
h = 40/200 = 0.20

1.00

	H (½)	h (½)
H (½)	HH (¼)	Hh (¼)
h (½)	Hh (¼)	hh (¼)

FIGURE 4-2 Probability Diagram

As long as these allele frequencies remain unchanged in a population whose members mate at random in respect to this trait, the genotype frequencies will appear in the same proportions, and the population will be in genetic, or Hardy-Weinberg, equilibrium. It follows that the recessive trait will not disappear when these necessary conditions are met, but will remain constant in frequency from one generation to the next.

One of the uses to which the Hardy-Weinberg principle can be applied is the estimation of the number of heterozygotes in a population for traits in which the heterozygote phenotype is indistinguishable from the phenotype produced by the homozygous genotype. If we assume that a certain population is in Hardy-Weinberg equilibrium, then the frequency of the three genotypes is $p^2 + 2\,pq + q^2$. If we find a population of 100 people in which 4 are recessive albinos (genotype *aa*), then some of the remaining 96 non-albinos must have the *AA* genotype and the remainder must have the genotype *Aa.* We can, using the Hardy-Weinberg principle, estimate how many of each kind of genotype of non-albinos are present, by calculating the allele frequencies for this locus and then placing these figures in the formula for genotype distributions.

If albinos constitute 4 percent of this population, the genotype *aa* frequency must be 0.04, or $q^2 = 0.04$. Since we need to know the allele frequency, q, we can determine this figure by taking the square root of the q^2 value: $\sqrt{0.04} = 0.20$. Remember that the two alleles at a single locus can never add up to more than 1.00; so if $q = 0.20$ then p must be 0.80. Thus we have estimated the allele frequencies as $p = 0.80$ and $q = 0.20$.

The next step is to insert these frequencies in the formula which describes genotype frequencies: p^2, the frequency of the homozygous genotype *AA*, is $(0.80)^2$, or 0.64; the frequency of the heterozygous genotype *AA* is $2 \times (0.80)(0.20)$, or 0.32. It follows from these calculations that the genotype frequencies of this population can be estimated as:

$$p^2_{(AA)} = 0.64 \qquad 2pq_{(Aa)} = 0.32 \qquad q^2_{(aa)} = 0.04.$$

Of course, we cannot tell from these operations which of the non-albinos is a heterozygote and which is a homozygote, but we can estimate the frequencies of the three genotypes in the whole population.

With these basic operations in mind, for the sake of experience consider the case of the mythical "Hairies," a population of 1000, of which 19 percent have hairy elbows—a trait previously identified among an Amish group and thought to be a dominant condition. If we want to estimate how many individuals are homozygous for this trait, all of whose children must therefore be affected, we need only follow the procedure already described, as set out in Example A.

EXAMPLE A

	Hairy Elbows	*Smooth Elbows*	*Total*
PHENOTYPE			
Number	190	810	1,000
Frequencies	.19	.81	1.00
GENOTYPE			
Frequencies	$(p^2 + 2pq)$	(q^2)	

CALCULATING ALLELE FREQUENCIES:

$$q = \sqrt{q^2} = \sqrt{.81} = .90$$

$$p = 1.00 - q = 1.00 - .90 = .10$$

ESTIMATING GENOTYPE DISTRIBUTIONS:

$$p^2{}_{AA} = (.10)^2 = .01$$

$$2pq = 2\,(.10)(.90) = .18$$

$$q^2 = (.90)^2 = .81$$

Of this group, only 10 individuals, 1 percent, are homozygous for the hairy elbow allele, while 180 are heterozygous at this locus.

This use of the Hardy-Weinberg principle is not limited to bi-allelic loci, but can be extended to genes for which any number of alleles is known. The *ABO* blood-group system has been studied in many populations, and estimations of the genotype distributions and allele frequencies are entirely feasible in this manner. This can be demonstrated by examining, for the moment, the distribution of only the three major alleles (A, B, and O) of this system, and using the symbols p, q, and r to refer to the frequency of the three alleles. The relationships of the phenotypes, genotypes, and genotype frequencies of each category are shown in Table 4-1. These genotype frequencies are derived from the formula shown earlier, expanded to include 3 alleles:

$$(p + q + r)^2 = p^2 + q^2 + r^2 + 2pq + 2pr + 2qr.$$

The problem in Example B can serve to illustrate how the procedure applies to a system in which the three alleles are present. We could now estimate that 34 percent of this population are heterozygous at the *ABO* locus since 2 percent of all individuals have *AB* blood type ($2pq$), 16 percent are *AO* genotype ($2pr$) and 16 percent ($2qr$) have the *BO* genotype.

If populations are in genetic equilibrium, then this technique is a nice toy, and you can calculate allele frequencies to estimate genotype frequencies which are identical to those from which you originally calculated allele frequencies. This sounds like a perfect wheel-spinning exercise, but it really isn't. In fact, the Hardy-Weinberg principle assumes conditions which must be very rarely met; so the technique is essentially used as a means of comparing reality to an ideal model, in order to detect if any equilibrium-disturbing forces *are* operating. Take the following example: a certain population of 200 people was tested for the *MN* blood-group phenotypes and was found to include 80 percent *M* blood type and 20 percent *MN* phenotypes. We can "solve" for allele frequencies and estimated genotype distributions in the same form already used (see Example C).

In this example, the population is *not* in genetic equilibrium, a finding which would titillate the soul of any investigator because this means the beginning of a search for explanations. Populations in genetic equilibrium are plainly a bore and unlikely to open up the kinds of questions and investigations which are the delight of the anthropologist. So, to start with

TABLE 4-1 ABO Blood Group System, Simplified

Phenotype	*Genotype*	*Genotype Frequencies*
A	AA	p^2
	AO	2pr
B	BB	q^2
	BO	2qr
AB	AB	2pq
O	OO	r^2

EXAMPLE B Calculating Allele and Estimated Genotype Frequencies at the ABO Locus

Blood Types:	*A*	*B*	*AB*	*O*	*Total*
PHENOTYPE					
Number	17	17	2	64	100
Frequencies	.17	.17	.02	.64	1.00
GENOTYPE					
Frequencies	$(p^2 + 2pr)$	$(q^2 + 2qr)$	$(2pq)$	(r^2)	

Calculating Allele Frequencies

$$r_O = \sqrt{r^2} = \sqrt{.64} = .80$$

$$p_A = \sqrt{A + O} - \sqrt{O} = \sqrt{.17 + .64} - \sqrt{.64} = \sqrt{.81} - \sqrt{.64} = .90 - .80 = .10$$

$$q_B = \sqrt{B + O} - \sqrt{O} = \sqrt{.17 + .64} - \sqrt{.64} = \sqrt{.81} - \sqrt{.64} = .90 - .80 = .10$$

Estimating Genotype Frequencies:

$$A = p^2 + 2pr = (.10)^2 + 2(.10)(.80) = .01 + .16 = .17$$

$$B = q^2 + 2qr = (.10)^2 + 2(.10)(.80) = .01 + .16 = .17$$

$$AB = 2pq = 2(.10)(.10) = .02$$

$$O = r^2 = (.80)^2 = .64$$

$$1.00$$

EXAMPLE C MN Blood Type Study

Blood Types:	*M*	*MN*	*N*	*Totals*
PHENOTYPES				
Number	160	40	0	200
Frequencies	.80	.20	.0	1.00
GENOTYPE				
Frequencies	p^2	$2pq$	q^2	

Calculating Allele Frequencies:

$$p = p^2 + pq = .80 + .20/2 = .80 + .10 = .90$$

$$q = q^2 + pq = 0 + .20/2 = 0 + .10 = .10$$

Estimating Genotype Frequencies

$$p^2{}_{MM} = (.90)^2 = .81$$

$$2pq_{MN} = 2(.90)(.10) = 2(.09) = .18$$

$$q^2{}_{(NN)} = (.10)^2 = .01$$

the obvious, findings of this kind would probably engender the following responses:

1. Recheck the testing procedures. Perhaps the anti-sera used to test the blood samples are faulty. It always helps in such cases if the anthropologist conducting such surveys is a heterozygote, since the anti-sera can be tested against the investigator's blood at fairly frequent intervals to ensure its accuracy. More eager investigators who are themselves not blessed with an appropriate blood type have been known to exploit spouse, children, even assistants, with appropriate blood types for just this purpose, raising the spectre of the successful homozygous researcher returning from his journeys with anemic, but heterozygous, families and associates. The advent of modern equipment, where it can be used in the field, virtually guarantees that technical problems are unlikely to produce the results seen here, but a review of technique in blood-typing procedures is another necessary condition in evaluating these kinds of results.

2. Is assortative mating for this trait characteristic of this population? Recall that the Hardy-Weinberg principle assumes random mating, so deviations in this respect can affect results—a point to be discussed at greater length in Chapter 5. Obviously, a competent investigator will want to collect the genealogies of the study population, in part for this very reason, and the researcher who fails to do so may overlook a significant aspect of genetic studies in human groups. In this case, however, it is unlikely that assortative mating in respect to this trait is an issue, since most people are totally unaware of their blood types for the *MN* antigen system and scarcely anyone is likely to marry or avoid marriage on the basis of compatibility in this respect.

3. How about sampling procedures? If I take blood samples from only 50 people out of a total population of 50,000, by chance alone I may well not have collected blood from any individuals with the *N* phenotype. In the present example every member of the population was tested, so this is not an issue here, but it is a highly critical consideration in research design, procedure, and success.

4. Are evolutionary forces operating which can explain these results? Here the investigator must utilize every available resource to try and detect what kinds of mechanisms could produce these results, usually without being able to definitely exclude all alternative possibilities. In this case, if the genealogical data confirmed that there had been no admixture of genetic materials from other populations and no migration, then gene flow would not be a tenable explanation. Mutation, always a rare event, could not explain the testing results. Thus we are left with two viable alternatives: genetic drift and natural selection.

If this population had been established for more than one generation, then an unrepresentative group of founders could not suffice as explanation of the observed genetic disequilibrium—since genetic equilibrium should be established within one generation regardless of the allele frequencies of the founder group. But random variations from one generation to the next (*intergenerational gamete sampling variation*) in a population of such small size is a viable explanation which cannot be rejected on the basis of the present evidence. The alternative hypothesis—of natural selection operating in favor of the heterozygote *MN*—is probably more satisfying and entirely consistent with the observed data. The observed population has an excess of heterozygotes (20 percent) over the expected value (18 percent) predicted by the genetic equilibrium model, and studies in a number of other populations have also found an excess of observed heterozygotes at this locus. This finding is so common as to suggest broad application, and the deviation noted in this population is in the exact direction predicted by such an explanation.

After all these hypothetical cases, it may be of some interest to see how this approach is used in an actual study, one which I carried out on the island of Saipan. In 1939, other investigators had collected blood samples from 678 Saipanese adults, with the results shown in Table 4-2. If the conditions stipulated by the Hardy-Weinberg principle were met, the next generation should have *ABO* blood group frequencies virtually identical to those reported in 1939. But, in 1964, exactly twenty-five years later, I collected information for 1057 Saipanese blood donors with the results shown in Table 4-3.

Clearly, this population was not in genetic equilibrium. There had been a great increase in the frequency of *B* blood type in the group, as well as a decrease in the frequency of the *OO* genotype. Blood typing results were obtained by competent investigators using entirely satis-

TABLE 4-2 ABO Blood Group Studies, Saipan, 1939

Phenotype Frequencies	A = 0.288	B = 0.128	AB = 0.062	O = 0.522
Allele Frequencies	p = 0.177, q = 0.084, r = 0.723			

TABLE 4-3 ABO Blood Group Studies, Saipan, 1964

Phenotype Frequencies:	A = .281	B = .201	AB = .046	O = .472
Allele Frequencies:	p = .181, q = .133, r = .687			

factory hospital and laboratory procedures. There is no evidence that nonrandom, or assortative, mating took place in respect to this trait. Sample size and sampling procedures were adequate and unbiased in respect to the various blood types. Mutation could not account for the rapid changes which took place in the twenty-five year period being considered, nor were selective pressures of the necessarily high intensities likely to have been experienced by any human group. Genetic drift cannot be completely excluded, but the native population of Saipan increased from 3,179 in 1939 to 8,404 in 1964. However, over 20,000 Japanese troops were stationed on Saipan from the 1930s until the end of World War II, and genealogies clearly indicated that considerable admixture had taken place between the natives and Japanese resident soldiers and civilians. The admixture of Japanese into the native population is entirely consistent with the observed changes, since *B* blood type is more common and *O* blood type less frequent in Japanese populations than in native groups of unmixed ancestry in this part of the world.

In summary, the Hardy-Weinberg principle provides both an ideal model against which real data can be measured and a powerful tool for the study of population genetics. From the basic formula, more sophisticated models have been derived which permit highly complex analyses of population genetics data, and recent advances have incorporated a probabilistic approach to problems in this area. In chapters 5 through 9 most of the topics already mentioned—assortative mating, genetic drift, natural selection, gene flow, mutation—will be considered at greater length, and with minimal mathematical accompaniments. But the basic concepts developed in this chapter are absolutely fundamental to the understanding of basic principles in population genetics and a broader appreciation of the nature of human variation.

PROBLEMS

---

1. A population was tested for *MN* antigens with the following results: $M = 49$, $MN = 42$, $N = 9$. What are the allele frequencies for this antigenic system?

   $p =$ ________, $q =$ ________

2. What are the allele frequencies if phenotype numbers are as follows: $M = 98$, $MN = 84$, $N = 18$?

   $p =$ ________, $q =$ ________

3. What are the allele frequencies if phenotype numbers are as follows: $M = 68$, $MN = 58$, $N = 12$?

   $p =$ ________, $q =$ ________

4. What are the allele frequencies if, in a population of 1300, 13 people are recessive albinos?

$p =$ ________, $q =$ ________

4-1) How many heterozygote genotypes are predictably present in this population?

*Aa*________

4-2) If all 13 recessive albinos die before reproducing, what would be the allele frequencies in the population after all albinos have died?

$p =$ ________, $q =$ ________

5. In a population of 100 people, the following *ABO* phenotypes were recorded: $O = 49$, $A = 32$, $B = 15$, $AB = 4$. How many heterozygotes are present in this population?

Heterozygotes________

6. In a population of 816, the following *ABO* phenotypes were recorded: $O = 400$, $A = 261$, $B = 122$, $AB = 33$. What is the frequency of the three alleles?

$p =$ ________, $q =$ ________, $r =$ _____

7. In a population of 1206 people, the allele frequencies for the *ABO* blood group frequencies are: $r = .8$, $p = .1$, $q = .1$. What are the genotype frequencies for the homozygotes in this population?

$x^2 =$ ________, $p^2 =$ ________, $q^2 =$ ________

7-1) How many heterozygotes are in this population?________

RECOMMENDED READINGS AND REFERENCES

CAVALLI-SFORZA, L.L. 1977 *Elements of Human Genetics.* 2nd ed. Menlo Park: W.A. Benjamin, Inc.

CROW, J.F. 1976 *Genetics Notes.* 7th ed. Minneapolis: Burgess Publishing Company.

LERNER, I.M., and W.J. LIBBY 1976 *Heredity, Evolution and Society.* 2nd ed. San Francisco: W.H. Freeman Co.

LI, C.C. 1955. *Population Genetics.* Chicago: University of Chicago Press.

MALECOT, G., ed. 1969. *The Mathematics of Heredity.* Translated by D.M. Yermanos. San Francisco: W.H. Freeman and Co.

METTLER, L.E., and T.G. GREGG 1969. *Population Genetics and Evolution.* Englewood Cliffs: Prentice-Hall, Inc.

MOLNAR, S. 1975. *Races, Types and Ethnic Groups.* Englewood Cliffs, N.J.: Prentice-Hall, Inc.

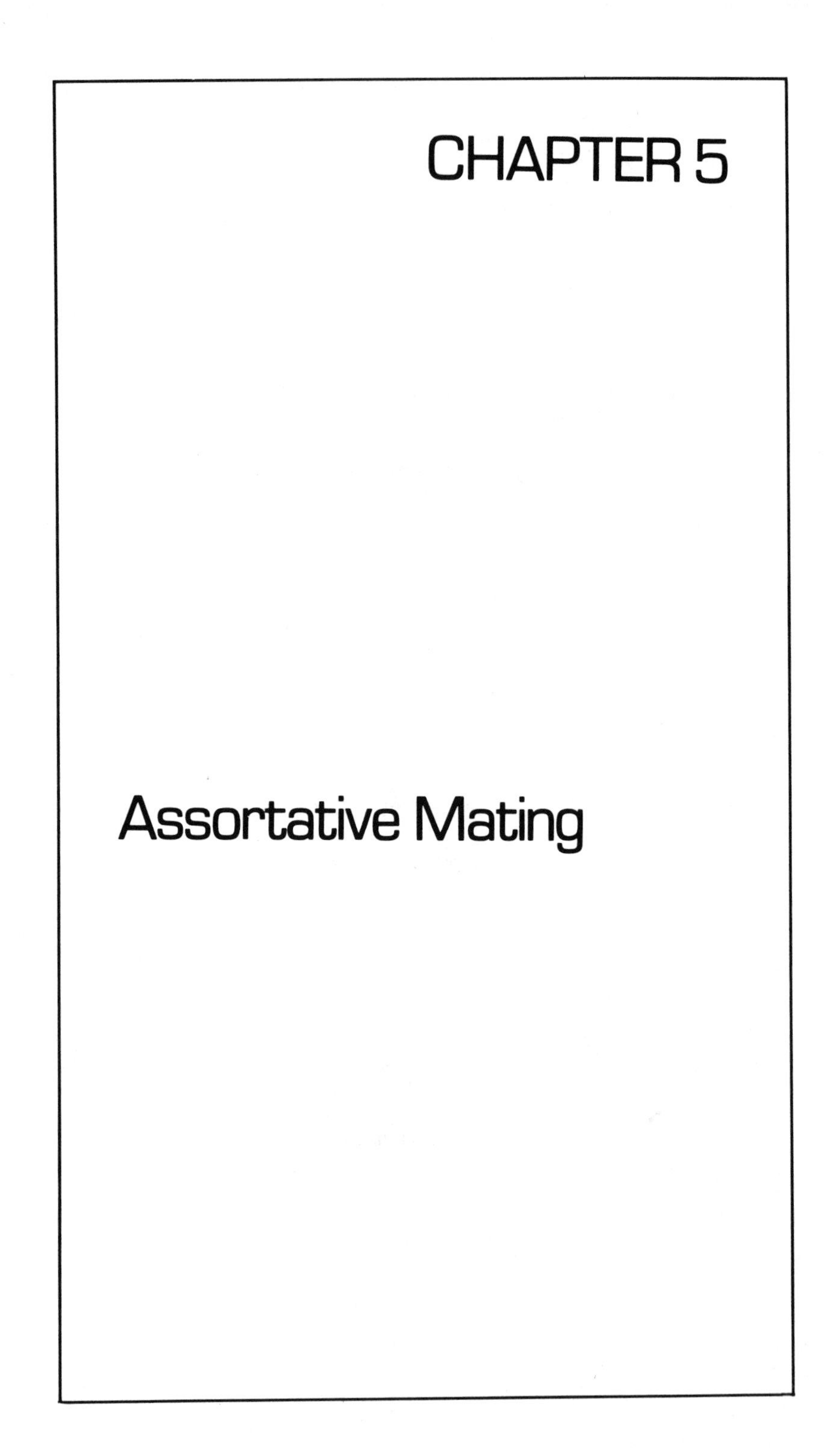

CHAPTER 5

Assortative Mating

Chimpanzee females do not engage in mating behavior with their male offspring, and male fruit flies under certain conditions exhibit preferential mating behavior with alien females; the human species, however, has developed patterns of non-random mating to a degree unparalleled in the animal world. *Assortative, or non-random, mating* occurs in every human society, thus violating the first required condition (panmixia) for the operation of the Hardy-Weinberg model in human populations. Since the bases of our varied and diverse mating proscriptions and prescriptions are primarily cultural, the subject of assortative mating provides a unique opportunity to examine the biocultural interactions which so deeply affect the origin and maintenance of human variation.

Not all forms of non-random mating involve traits with a direct genetic basis. Various studies have shown that propinquity, or physical proximity, plays an important role in mate selection in many human societies. Unless you are a member of the international jet set, your future spouse is likely to be a neighbor. Even in highly mobile societies, a young adult has a high probability of marrying someone who lives at that time within a radius of a few miles from his or her own residence. Since neighborhoods often include people of similar religious, ethnic, social, or economic characteristics, especially in large cities, the propinquity factor may result in a high degree of sociological assortment; but any genetic consequences are incidental to the selected factor of distance. The chances are that you may very well marry the boy or girl, if not next door, at least not too many doors away.

Similarly, high correlations between age of mates have been demonstrated in a number of studies. In the United States and western Europe, husbands tend to be several years older than their wives. While this has little, if any, influence on the genetics of these populations, it does result in some interesting demographic consequences. At present, women in most Western societies have a longer life expectancy than males, so our marital age patterns virtually guarantee that many elderly women will spend an appreciable portion of their final years as widows. Sensibly, since a woman's prime years for childbearing occur during her twenties, young college women should be seeking mates on the playgrounds of the nearest junior high schools. The great social, personal, and economic burdens of widowhood would be alleviated by such a change, but it is unlikely that entrenched values and attitudes will so readily adjust to this dilemma.

When members of a population who are similar in respect to a genetically based phenotypic trait mate with one another more often than would happen by chance alone, *positive phenotypic assortative mating* (*homogamy*) occurs. Since few of us are equipped to determine similarities in biochemical traits or to make fine metric distinctions, assortative matings usually involve gross and readily perceived differences, such as that involved in homogamous matings for stature observed in the United States and Britain. *Negative phenotypic assortative mating, heterogamy,* likewise involves readily detectable traits. It has been claimed that persons with red hair tend to mate preferentially with individuals of all other hair colors and to avoid matings with others who share the same red hair color; among the San Blas Indians of Panama, albinos mate less often with other albinos than would be predicted if mating were random in respect to this trait.

The predictable genetic consequences of phenotypic assortative mating can be considered through the use of a simplified and hypothetical example. Assume that we are dealing with an autosomal recessive trait, such as "sweaty-feet disease," caused by a metabolic disorder which results in the formation of isovaleric acid, and imagine a population in which:

(1) the allele frequencies are $S = .60$ and $s = .40$; and

(2) the genotype distributions are $SS = .36$, $Ss = .48$ and $ss = .16$.

In a stable population of 100 people, with random mating and no fertility differentials by genotype, the F_1 population would again consist of 84 people with normal-smelling feet and 16 people afflicted with smelly-feet disease:

$p^2 = .60^2 = .36\ (100)$	36 *SS*	
$2pq = 2(.60)(.40) = .48\ (100)$	48 *Ss*	
		84 "normal"
$q^2 = .40^2 = .16\ (100)$	16 *ss*	16 "Sweaty-feet"

(Hardy-Weinberg equilibrium)

However, if victims of this disorder mated only with fellow-sufferers of the disease, while the remainder of the populaton mated at random, what would the results be? To answer this question, it is useful to separate the total population into two subgroups, one consisting only of the 16 *ss* individuals who will be mating with each other and producing 16 *ss* children, and to place in the other group the remainder of the population, all with normal-smelling feet, but including the genotypes *SS* and *Ss*, as illustrated in Table 5–1.

Clearly, the results of phenotypic assortative mating are to change genotype distributions without any change in allele frequencies. Alleles are reassorted into different genotype distributions, but the allele frequencies remain unaltered. This may, indeed, produce different conditions for the operation of natural selection as more autosomal homozygous combinations appear in the population, but, in and of itself, assortative mating has the single effect of altering genotype distributions.

A second major form of non-random mating, termed *genotypic assortative mating,* or *inbreeding,* involves preferential mating between individuals who are genetically related to each other and therefore have a certain probability—dependent on the degree of relationship—of sharing a gene common by descent from a shared ancestor. It is important to note here the distinction between inbreeding and *incest;* the latter term refers to prohibitions against mating between individuals who are culturally defined as belonging to a proscribed category of potential mates. In many cases, incest restrictions do refer to individuals who are genetically related to *Ego* (the centrum of a kinship terminological system), but non-genetically related individuals may also be included in such a forbidden category. For example, Georgia and Tennessee have both had laws prohibiting the marriage of male Ego and his son's wife; Virginia has banned the marriage of a man and his wife's stepdaughter. Obviously, these individuals have no necessary genetic relationship to male Ego; the prohibition of marriage between such individuals does not prevent

TABLE 5-1 Calculation of Allele Frequencies under Conditions of Phenotypic Assortative Mating

	Sub-group I				*Sub-group II*			
Genotype	SS	Ss	ss	Σ	SS	Ss	ss	Σ
number	36	48	0	84	0	0	16	16
frequency	.429	.571	0	1.00	0	0	1.00	1.00
Allele frequency	p = .429 + .2855 = .7145				p = 0.0			
	q = 0 + .2855 = .2855				q = 1.00			

F_1 generation:

$p^2 = (.7145)^2 = .51\ (84) = 43$ SS

$2pq = .41\ (84) = 34$ Ss

$q^2 = .08\ (84) = 7$ ss; total 84

$q^2 = 1.0^2 = 1.0\ (16) = 16$ ss; total 16

F_1 TOTALS:

	SS	Ss	ss	Σ
Genotype number	43	34	23	100
frequency	.43	.34	.23	1.00

Allele frequency:

p = .43 + .17 = .60 S

q = .23 + .17 = .40 s

inbreeding, but bans the marriage of persons belonging to certain culturally proscribed categories of social kin.

Conversely, many societies practice forms of preferred matings between socially defined kin who, again, may or may not be genetically related. During Biblical times, a woman was encouraged to marry her dead husband's brother (*levirate*); in many extant societies, a man is expected to seek additional wives among his wife's sisters (*sororate*). In neither custom is any necessary genetic relationship involved.

Preferred mating between genetically related individuals has been practiced in many societies, although there is great variation in the categories of prescribed and proscribed mates. Virtually all known societies have discountenanced marriage between siblings, although such matings were encouraged in the royal lines or noble families of native Hawaii, ancient Egypt, and pre-Contact Peru. Since siblings have a probability of 1/2 shared genes by virtue of descent from common parents, the offspring of a sibling mating is said to have a *coefficient of inbreeding,* or

F, of 1/4 (i.e., $1/2^2$). F refers to the probability that both alleles at a locus are received from an identical ancestral source; it ranges from a value of 1.0 in the case of self-fertilized organisms, such as self-fertilized plants, to a value of 0.0 where no inbreeding has occurred. The same symbol is used to indicate the degree of inbreeding in a population in comparison to a panmictic population in which F would have a value of 0.0.

More commonly, inbreeding in human populations involves matings between cousins, the offspring of a sibling pair, each of whom is married to another individual. In a number of Middle East societies, a man is encouraged to marry his father's brother's daughter, a parallel cousin; among the Trobriand Islanders, a man's preferred mate was his father's sister's daughter, a cross-cousin. Of course, such ideals were not always met in these societies, but the positive value attached to such practices encouraged a higher rate of genotypic assortative matings than would have occurred otherwise. In the case of first-cousin matings, the coefficient

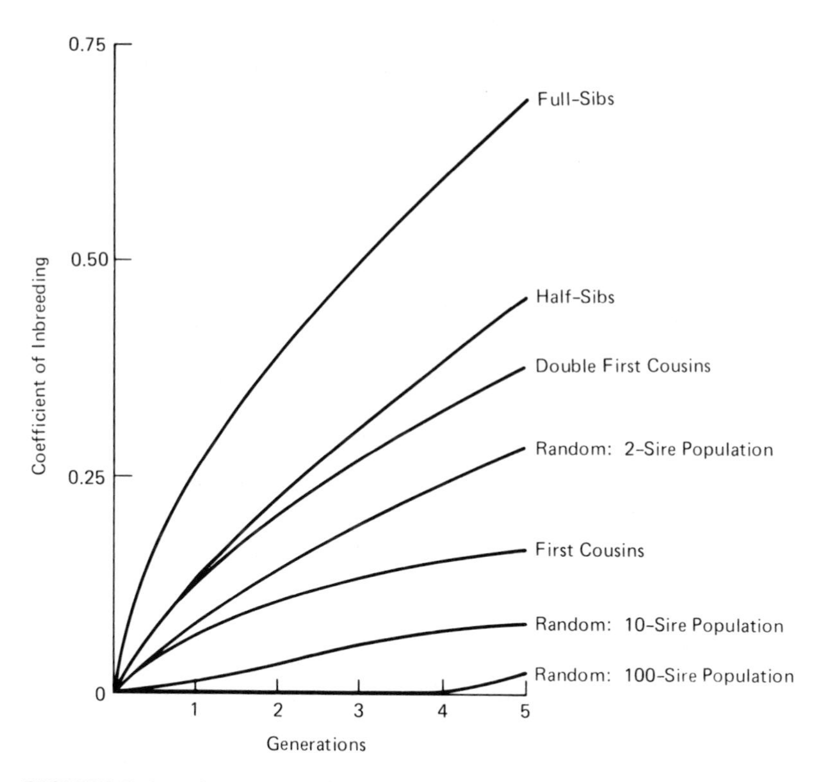

FIGURE 5-1 Increase in Homozygosity under Inbreeding

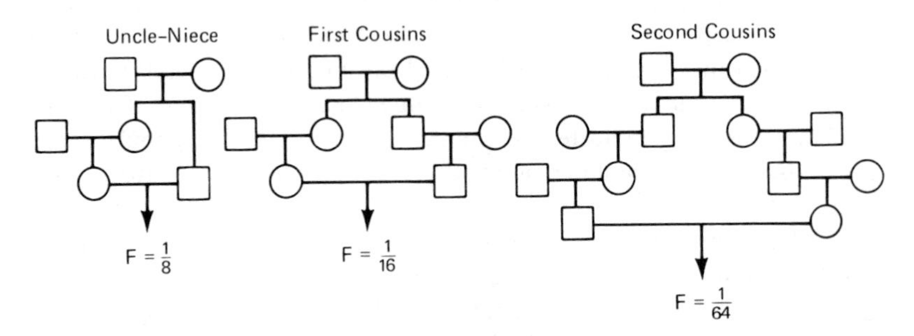

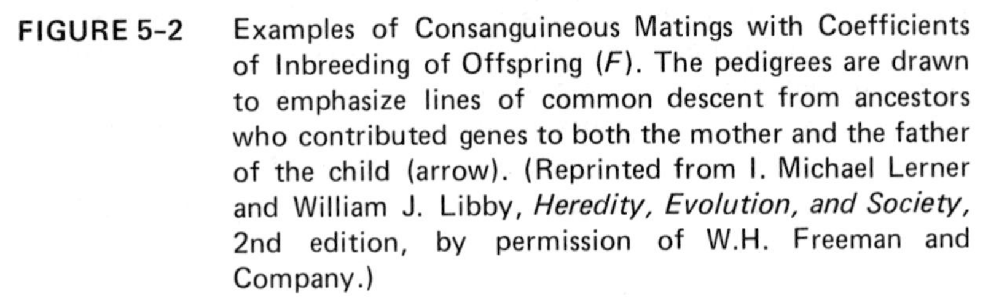

FIGURE 5-2 Examples of Consanguineous Matings with Coefficients of Inbreeding of Offspring (F). The pedigrees are drawn to emphasize lines of common descent from ancestors who contributed genes to both the mother and the father of the child (arrow). (Reprinted from I. Michael Lerner and William J. Libby, *Heredity, Evolution, and Society,* 2nd edition, by permission of W.H. Freeman and Company.)

of inbreeding for autosomal genes is correspondingly lower (F = 1/16, or $1/4^2$) than for sibling matings, but represents a distinct increase in homozygosity above the levels expected in a panmictic population.

Another hypothetical example may help to illustrate the predicted genetic consequences of inbreeding in human populations. Imagine a native island population from which the allele for recessive albinism is wholly absent, and, therefore, in which all individuals are of the genotype *AA*. Assume further that this native group is visited by an albino seafarer who mates with 20 native women, producing one child by each woman, while the remaining natives mate at random among themselves, producing a total of 80 children. Thus, in the F_1 generation, 80 individuals are of genotype *AA* while 20 have the *Aa* genotype (see Table 5-2). The allele frequencies in this F_1 population are A = .90 and a = .10. If random mating then prevails, the predicted genotype distributions for the F_2 and subsequent generations would be *AA* = .81, *Aa* = .18, and *aa* = .01.

Suppose, however, that the descendants of the seafarer mated only with each other (inbreeding), while random mating characterized the rest

TABLE 5-2

Genotype	*AA*	*Aa*	*aa*	Σ
number	80	20	0	100
frequency	.80	.20	0.0	1.00

TABLE 5-3 Calculation of Allele Frequencies under Conditions of Genotypic Assortative Mating

	Group I		*Group II*	
Matings:				
	AA x AA = 1.00 (80) = 80 AA		Aa x Aa = 1/4 AA (20) = 5 AA	
			1/2 Aa (20) = 10 Aa	
			1/4 aa (20) = 5 aa	

Total F_2 Population:

Genotype	*AA*	*Aa*	*aa*	Σ
number	80 + 5	0 + 10	0 + 5	100
frequency	.85	.10	.05	1.00
Allele frequency:		p = .85 + .05 = .90		
		q = .05 + .05 = .10		

of the population. In this case, the F_2 generation would exhibit different genotype distributions, although the allele frequencies would remain $A = .90$ and $a = .10$. Again, genotype distributions are changed without altering allele frequencies; but if random mating is then resumed, these effects are reversed in one generation of mating and the F_3 generation genotype distributions would revert to: $AA = .81$ ($.90^2$); $Aa = .18$ ($2 \times .90 \times .10$); and $aa = .01$ ($.10^2$).

A good deal of folklore exists and a considerable amount of research effort has been expended on the subject of the overall effects of inbreeding. It has been argued by some writers that the universality of incest regulations expresses an overt or instinctive awareness of the deleterious consequences of inbreeding. However, as noted, incest regulations do not pertain exclusively to matings of genetic relatives; so, more sophisticated explanations of the origins of incest restrictions are required. Certainly, the increased autosomal homozygosity which results from inbreeding is of relevance, primarily in respect to deleterious traits expressed in the homozygous recessive phenotype. A completely dominant allele with lethal effects is subject to natural selection when present in either the heterozygous or homozygous condition.

Inbreeding can occur at either the pedigree level (*pedigree inbreeding*) or at the level of the population. *Population inbreeding* includes the summed, relative proportion of all pedigree inbreeding and/or (in the case of small-scale populations) the unintentional inbreeding of individuals whose genetic relationship is an inherent function of the limited number of possible ancestors in a group of limited size and restricted ancestral origins. Even though pedigree inbreeding is assiduously avoided, a considerable amount of mating between remotely related individuals may occur, especially when the living population is descended from only a few reproductively effective founders. A number of studies in several religious isolates, the Amish, Dunkers, and Hutterities, has revealed a comparatively high incidence of genetic defects produced by homozygous recessive genotypes, and it has long been observed that *consanguinity* (genetic relatedness) of parents is common in persons exhibiting rare hereditary conditions.

Roberts has found an inverse relationship between intelligence quotient score and F value. Schreider recorded a significant negative correlation between stature and the coefficient of inbreeding in seventy departments of France. Studies in Japan, where the coefficient of inbreeding is elevated by preferential cousin mating, have shown increased rates of infant mortality and congenital abnormalities. Studies in France, Sweden, the United States, and Japan have shown increased frequencies of certain physical diseases and mental disorders among children of first-cousin mating.

Since so many studies demonstrate the detrimental consequences of inbreeding, the well-being of any human population would seemingly be advanced by an immediate surcease in genotypic assortative mating. But this is no simple matter. Inbreeding merely increases homozygosity and provides more homozygous recessive phenotypes on which natural selection can operate. The removal by natural selection of recessive alleles which produce deleterious traits when present in the homozygous condition alters the gene pool of a population in the direction of an increased frequency of the favored allele. In a panmictic population, deleterious recessive alleles would be present more often in heterozygous combination and, thus, protected from the operation of natural selection.

In fact, some research has not supported the simplistic view that inbreeding is inherently or inevitably deleterious. Increased fertility from inbred matings has been reported in several studies. In Japan, Schull found a lower rate of harelip in inbred matings than in matings of unrelated parents; Neel has presented statistics showing that in Sweden more deaths occur before reproductive ages to children of unrelated spouses than to offspring of first-cousin marriages.

Two very closely inbred pedigrees reported in the literature have also failed to indicate detrimental effects of inbreeding. In the first instance, the critical mating reported by Birdsell involved that between an Afghan camel-driver and his own daughter by an Australian aborigine woman. The second case involves a complex pedigree which I recorded on the island of Yap and which culminated in a mating that produced a child whose coefficient of inbreeding has a value of 0.375. One reasonable explanation of such findings postulates that prolonged inbreeding over many generations in small populations has already resulted in the removal through natural selection of deleterious recessive alleles which have been repeatedly expressed in homozygous combinations. Indirect support of this interpretation is provided by the relatively recent appearance (within the last few hundred years) of those same religious isolates in which deleterious recessive conditions have been identified at such high frequencies. In a sense, panmixia delays the operation of natural selection by concealing deleterious alleles in the heterozygous state. But the allele which is disadvantageous in the present environment may, at some later time and under different conditions, prove selectively advantageous.

The effects of assortative mating are not merely a matter of academic interest. Current programs in genetic screening and prevention of certain hereditary diseases, notably sickle cell anemia and Tay-Sachs disease, in part seek to encourage selective matings of unaffected homozygous normals and heterozygote carriers. Sickle cell anemia, involving a hemoglobin variant which results in severe anemia in the homozygous condition, is considered a major health problem among American Blacks. Since affected persons rarely survive to reproductive ages, the condition normally occurs only as a result of matings between heterozygous parents

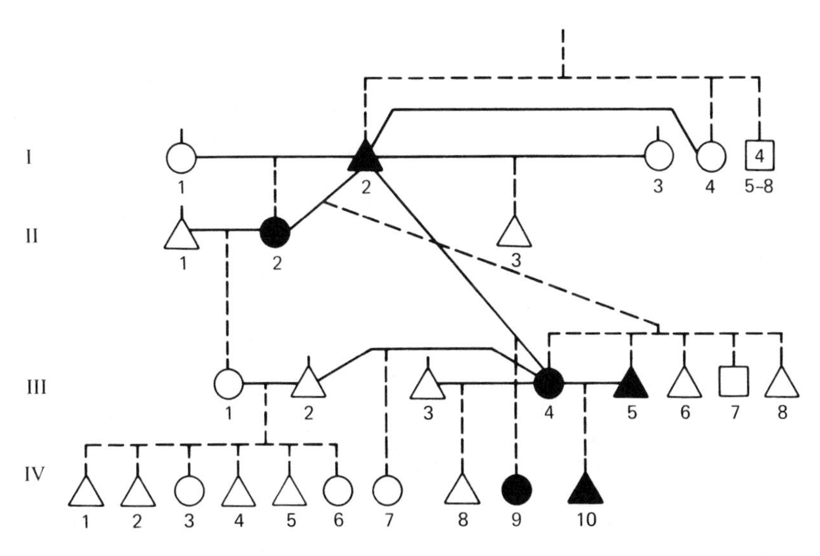

FIGURE 5-3 Pedigree Inbreeding (adapted from Underwood, 1977).

who have a milder form of disorder, sickle trait. These heterozygous individuals produce both normal and sickle hemoglobin and usually do not experience any debilitating symptoms. The heterozygotes experience less severe infestations of *Plasmodium falciparum*, the causative organisms of falciparum malaria, and in the past such individuals may have had a selective advantage in areas where this form of malaria was endemic. Falciparum malaria is no longer a health problem in the United States, so there is no known selective advantage for the heterozygote carrier of the sickle hemoglobin allele, while the homozygous offspring of two carriers is severely disadvantaged. Governmental support of programs to detect carriers and provide genetic counseling to prospective parents has the meritorious goal of attempting to reduce the anguish of producing diseased children and the personal and societal burden of caring for the afflicted. But discouraging the mating of heterozygous carriers, alone, will not lead to the removal of the allele from the gene pool.

Similarly, screening programs have begun in a number of large cities in an attempt to identify carriers of Tay-Sachs disease. This disease, also known as infantile amaurotic idiocy, involves a progressive deterioration of cerebral function which leads to death in early childhood. The disease is more prevalent among Jews than non-Jews, and more common among Ashkenazi Jews than other Jewish groups. The afflicted victim dies before reaching reproductive age, so the majority of cases (excluding newly arisen mutations) occurs as a result of the mating of unaffected heterozygote carriers of the allele. Assortative mating will not, of itself, remove the deleterious allele from the population. However, as techniques for the detection of this and other hereditary diseases from samples of the amniotic fluid surrounding the fetus are developed, a prospective mother can undergo *amniocentesis,* the sampling of fluids surrounding the unborn fetus. If the results indicate that the mother is carrying an affected child, she may elect to have an abortion performed. In this case, elective abortion would constitute a form of prenatal selection.

A third form of assortative mating, *sexual selection,* was described by Darwin as a "struggle" between members of one sex for the possession of members of the other sex. As he explained, "struggle" might be a matter of peaceful competition, even the display of plumage in the case of birds, or any means by which differential access to breeding partners was secured. In complex human societies, sexual selection probably has more often involved the differential access of males who are socially, politically, or economically advantaged to females endowed with valued characteristics. "Gentlemen prefer blondes," were this not circumvented by cosmetic means, might well result in a disproportionate number of women with darker hair color remaining without sexual partners, at least for a portion of their reproductive age period. On the other hand, it could be expected that males, regardless of hair color, but endowed with preferred biological or social attributes, would be relatively more successful in obtaining blonde partners.

As example of the operation of sexual selection has been furnished by Hulse's studies (1967) in Japan. Light skin color has been highly valued among the Japanese since long before Western contact took place, as revealed in ancient paintings and traditional literary masterpieces. This esthetic value has been expressed in marital preferences and practices, especially among males of the upper classes, with the result that a differential distribution of polygenic alleles for lighter skin color has taken place. Since socioeconomic classes are not completely reproductively isolated, class mobility has precluded a rigid stratification. But Hulse was able to find a significantly higher frequency of light skin color among the upper-class Japanese he examined. If this mating behavior continued more rigorously and over a much longer time period, it would predictably result in classes, not only distinguishable in terms of socioeconomic factors, but visibly distinctive in skin color.

A somewhat different mode of sexual selection involves the access of males to a larger number of females as a consequence of the status of such men. In many societies in which *polygyny,* or the simultaneous marriage of a man to two or more wives, occurs, such men usually comprise only a minority of the adult male population but are frequently the economically, socially or politically advantaged members of the society. Among the Yanomama of South America, two village headmen had sired twenty-eight children between them, or one-fourth of the entire village population. These village headmen, by virtue of their status, had greater access to Yanomamo females in polygynous unions, and also to captured females from other tribes. As a result, the contribution of these men to the population gene pool is much greater than that of ordinary adult male villagers. This phenomenon has been described as a form of genetic drift, but the mechanism of differential access accords with Darwin's description of sexual selection.

Assortative matings are the rule rather than the exception in human populations. This fact does not preclude the use of the Hardy-Weinberg model in studying the genetics of human groups, but provides a useful tool for interpreting some of the results of field research in human microevolution and for developing more sophisticated methods of analysis. Weinberg and, later, Dahlberg developed methods of estimating the frequency of cousin marriages in a European population from the proportions of recessive homozygotes. Subsequently, Dahlberg devised a useful method for estimating the size of a genetic isolate from the frequency of consanguineous unions.

Increased homozygosity should alert the investigator to the possibility of genotypic or phenotypic assortative mating patterns. This, in turn, should lead to the collection of extensive genealogical data and to the elicitation of cultural norms and esthetic standards which may be influencing marital partner choices. Truly, the genetics of a human population cannot be divorced from a study of the culture of a society, and this extends to past as well as present behavior. Wiesenfeld and Gajdusek (1976, p. 188) have most effectively summarized the need for an anthropological perspective:

> Careful collection of ethnohistory and anthropological data over long periods of time may prove as valuable as genetic surveys in unravelling the nature of human heterozygosity.

The direct genetic effect of genotypic and positive phenotypic assortative mating is limited to the redistribution of alleles into genotypes, with the result that homozygosity is increased. The immediate distributional effect is reversible in one generation if random mating is resumed. As a result of autosomal genotype reassortment, increased proportions of homozygotes are exposed to the operation of natural selection, which operates primarily against recessive alleles producing a deleterious phenotype when present in the homozygous condition.

The effects of sexual selection are less transitory, since these involve the differential contribution of some individuals to the gene pool of succeeding generations, thus paralleling the results of natural selection. Animal breeders have practiced selective breeding for many generations, attempting to produce domestic varieties in which preferred traits are more frequently expressed. However, no human society has rigorously enforced comparable selective breeding practices or maintained sufficiently rigid preferential patterns over enough generations to attain the degree of homozygosity which some domesticated animal breeds exhibit. In the future, human populations will begin to extend selective mating patterns to biochemical traits as these become more readily detectable. In combination with changing selective pressures, some of the rich genetic variation of the human species may be endangered. The long-term evolutionary consequences, no less than the immediate benefits, of the choices to be made demand our careful consideration and understanding.

PROBLEM: ASSORTATIVE MATING

An itinerant and lecherous trader, who was a recessive albino, landed on a remote island whose native population was totally lacking in the recessive albinism allele. The trader mated with 10 native women, producing one child by each woman.

1. What are the phenotype frequencies for the F_1 population ($N = 100$)?
   ________non-albinos ________albinos
2. What are the allele frequencies for the F_1 population ($N = 100$)?
   ________*A* ________*a*

Assume random mating takes place:

3. What are the genotype frequencies for the F_2 population ($N = 100$)?
   ________*AA* ________*Aa* ________*aa*
4. What are the allele frequencies for the F_2 population ($N = 100$)?
   ________*A* ________*a*

Assume that, for the F_1 population, the progeny of the trader may mate only with other progeny of the trader, while the remaining members of the F_1 generation mate at random with each other:

5. What are the genotype frequencies for the F_2 population ($N = 100$)?
   ________*AA* ________*Aa* ________*aa*
6. What are the allele frequencies for this F_2 population ($N = 100$)?
   ________*A* ________*a*

REFERENCES AND RECOMMENDED READINGS

CHAGNON, N.A.; J.V. NEEL; L. WEITKAMP; H. GERSHOWITZ; and M. AYRES. 1970. The influence of cultural factors on the demography and pattern of gene flow from the Makiritare to the Yanomama Indians. *American Journal of Physical Anthropology* 32:339–350.

DAHLBERG, G. 1947. *Mathematical Methods for Population Genetics.* New York: Karger

DARWIN, C. 1958. *The Origin of Species.* (Mentor edition) New York: New American Library of World Literature, Inc.

GLASS, B.; M.S. SACKS; E.H. JOHN; and C. HESS. 1952. Genetic drift in a religious isolate. *American Naturalist* 86:145–159.

HULSE, F.S. 1967. Selection for skin color among the Japanese. *American Journal of Physical Anthropology* 27:143–155.

LERNER, I.M. and W.J. LIBBY. 1976. *Heredity, Evolution and Society.* 2nd ed. San Francisco: W.H. Freeman and Co.

McKUSICK, V.A. 1964. Distribution of certain genes in the Old Order Amish. *Cold Spring Harbor Symposia in Quantitative Biology,* 29:99–115.

McQUEEN, D.V. 1975. Social aspects of genetic screening for Tay-Sachs disease: the pilot community screening program in Baltimore and Washington. *Social Biology* 22:125–133.

NEEL, J.V.; M. KODANI; R. BREWER; and R.C. ANDERSON 1949. The incidence of consanguineous matings in Japan. *American Journal of Human Genetics* 1:156–178.

ROBERTS, D.F. 1968. Genetic fitness in a colonizing human population. *Human Biology* 40:494–507.

SCHREIDER, E. 1969. Inbreeding, biological and mental variations in France. *American Journal of Physical Anthropology* 30:215–220.

STEINBERG, A.G.; H.K. BLEIBTREU; T.W. KURCZYNSKI; A.O. MARTIN; and E.M. KURCZYNSKI 1967. Genetic studies in an inbred human isolate. In *Proceedings of the Third International Congress of Human Genetics* (J.F. Crow and J.V. Neel, eds.). Baltimore: Johns Hopkins Press.

STERN, C. 1973. *Principles of Human Genetics.* 3rd ed. San Francisco: W.H. Freeman and Co.

UNDERWOOD, J.H. 1977. Inbreeding. In *The Perception of Evolution.* L.L. Mai, ed. *Anthropology UCLA*, Vol. 7 (1–2).

WIESENFELD, S.L., and D.C. GAJDUSEK. 1976. Genetic structure and heterozygosity in the kuru region, Eastern Highlands of New Guinea. *American Journal of Physical Anthropology* 45:177–190. Courtesy of Wisrar Institute Press.

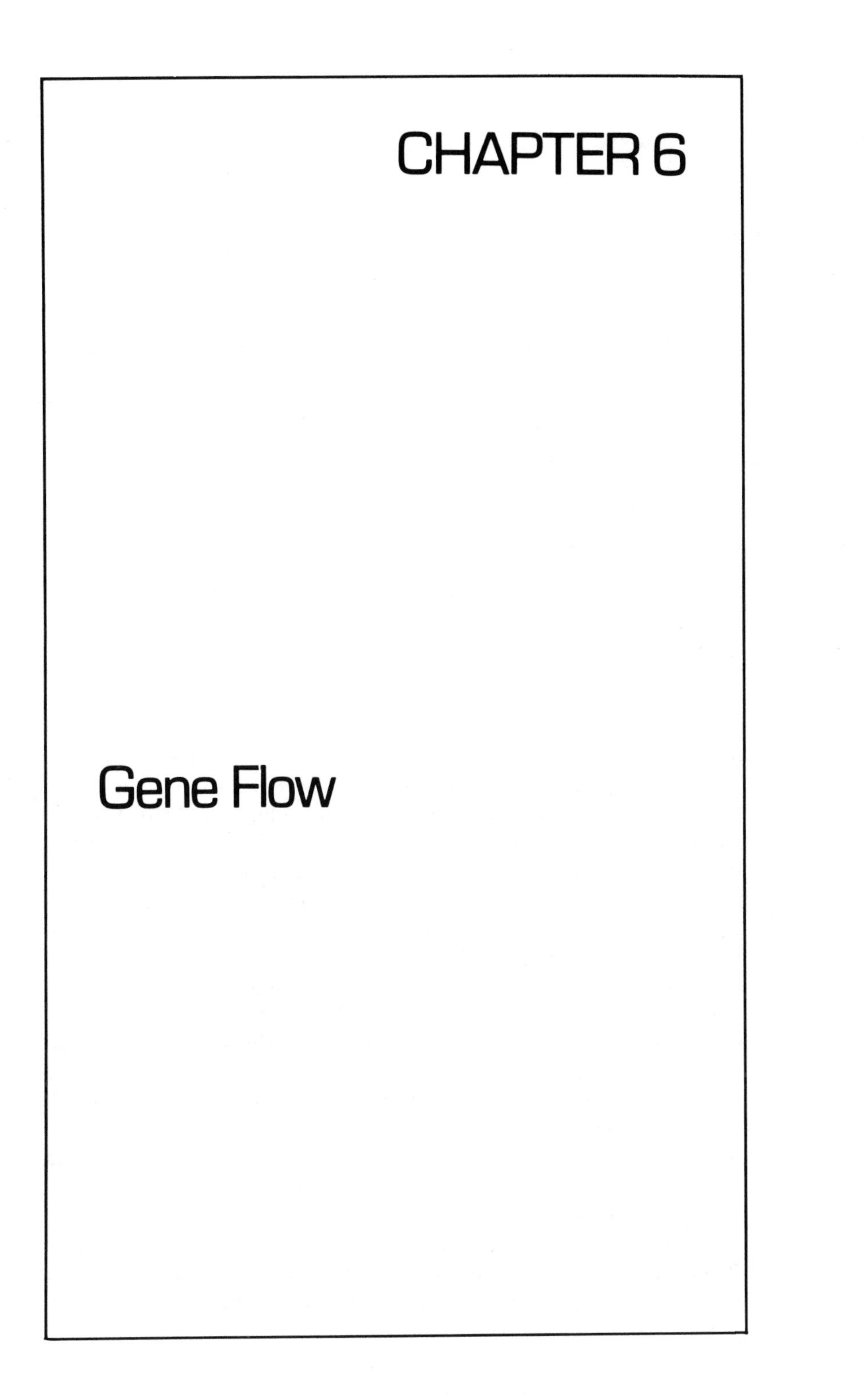

CHAPTER 6

Gene Flow

Gene flow, in the broadest sense, includes *intermixture,* the exchange of genes between populations, and *migration,* the movement of a population (and, necessarily, its gene pool) in space. Formerly, intermixture between human groups must have occurred primarily between members of neighboring populations, but, in our modern, mobile world, geographic barriers have been replaced as genetic isolating factors by cultural boundaries which can at least reduce genetic exchange between members of populations residing in a single community.

The Jewish colony of Rome provides but one of many possible examples of a population isolated by cultural factors from the larger community of which it is a part. Although colonies of Jews from Palestine had been living in Italy since at least 160 B.C., it was not until 1554 that the walled ghetto of Rome was closed by Pope Paul IV. In the ensuing 300 years, restrictive laws limited social intercourse and banned intermarriage of Jews and Christians in the Eternal City. Although the ghetto was opened in 1870, the ghetto area continues to be a center for Jewish social and religious life in Rome. Recent studies of historical records show that an *endogamous,* or in-marrying, Roman Jewish group continues to exist in Rome and differs in a number of genetic traits from the Italian Catholic population among which it lives. For example, the proportion of Roman

Jews with blood type *B* is more than twice as great (27 percent) as in the Italian Catholic population (10 to 11 percent). Intermixture, had it occurred, would have reduced this difference.

Intermixture reduces differences between formerly distinctive populations, thus posing the question of the origin and maintenance of variation in a *polytypic* species—a species comprised of phenotypically and genetically dissimilar populations. In most animal species, such diversity is commonplace. Localized populations of a species in geographic isolation may develop biological isolating mechanisms in response to the distinctive evolutionary histories of the separated groups, and may eventually form new species. This process is thought to have taken place in the evolution, in geographic isolation, of the distinctive New World and Old World monkeys from a common, more generalized, simian ancestor.

Localized human populations have developed genetic differences from other groups, but judging from the archeological evidence, migration has long characterized our species and prevented the absolute geographical isolation over lengthy periods which many other animal populations have experienced. Then, too, natural selection has been modified by technological advances in human groups, so that genetic differences have not always been a necessary condition of survival in disparate habitats. Clothing and housing, rather than the extensive necessary genetic changes for thick body hair, hibernation, or other biological adaptive techniques, have provided the Eskimo with a protective microenvironment not available to other mammals of the Far North.

In the past, relatively isolated human populations, living in different habitats, were subject to a variety of selective pressures. Mutations, always rare events, must have differed from one group to another, while the effects of random genetic drift could only have furthered the differentiation of these populations. What happens to human populations which have attained some degree of genetic adaptation when they migrate to different environments? What happens to the gene pools of such populations, each adapted to different conditions and with unique evolutionary histories, when intermixture takes place?

Even in the days when scholars believed different races were the products of separate, immutable creation, intermixture was an observable phenomenon which could not well be denied. Ignorant of the origins of human diversity, mixture between erroneously defined "pure" races was often considered an abomination, or, at best, was considered, by analogy to the production of sterile mules from matings between horses and asses, to be inherently limiting. Linnaeus's recognition of the common species membership of all living human races was followed by a host of racial classifications by Blumenbach, Retzius, Deniker, and others, which acknowledged the existence of a single polytypic human species, *Homo sapiens.* Neo-evolutionary theory, based on countless studies, successfully challenged false assumptions about the supposed purity of human races or the hypothesized dire consequences of intermixture, although racists continue to espouse such outmoded views even today.

Similarly, older views about the supposedly injurious effects of migration have largely succumbed to evidence that modification of behavior enables most humans to thrive in radically varied conditions. It is even amusing today to read that, in the nineteenth century, missionary ladies departing for the Pacific were warned of the debilitating effects of tropical climates at the same time that they were admonished to bring along adequate numbers of flannel petticoats for their tours of duty abroad. Today we know that American troops sent to desert regions of Africa and the tropical Pacific in World War II, protected by innoculations and prophylactic measures from endemic diseases and parasitic infestations, did not experience the chronic lassitude and fatigue reported by earlier European invaders of these areas. And the experiences of World War II and the building of the Alaskan pipeline showed that sensible precautions about exposure and the use of protective clothing enable the American black to be fully effective in the Arctic despite evidence of physiological differences in cold response.

In the New World, containing so many migrant populations as well as a rich assortment of groups formed from the intermixtures of native Indians, European settlers, and transported Africans, opportunities for the study of gene flow in human populations were readily available. Studies of the effects of migration can be said to have begun in the United States with the publication in 1911 of Boas's investigations of the differences between migrants and their American-born children. Subsequent studies in Hawaiian-born Japanese, Chinese born in the United States, migrant Mexican-Americans and their children, and among various other groups, confirmed the existence of characteristic differences in first-generation offspring in such traits as head shape, body weight, body height, and pigmentation.

Many investigators ascribed these changes in height and weight of the migrant offspring generation to improved nutritional conditions in the new environment, since it had been noted for some time that nutritional stress had retarding effects on the growth of children and on their ultimate achieved adult body size. It had also been repeatedly observed that adults in lower economic levels (presumably with poorer diets) of complex societies were characteristically smaller in body size than their more affluent peers. Even the *secular trend* in stature (an increase in average height between generations), recorded since at least the turn of the century in

European, American and, more recently, in Japanese populations, was thought to reflect, at least in part, improving nutritional standards over time. Another secular trend, toward decreasing age of *menarche* (the beginning of menstruation), was also thought to reflect the acceleration of growth processes that accompany improved nutrition. In 1971 Frisch and Revelle concluded that menarche was triggered by metabolic shifts which occurred only after a critical body fat to body mass relationship was attained. Thus, an acceleration of prepubescent growth processes due to improved nutrition should lead to an earlier average age at menarche.

What more attractive explanation could be found than that which ascribed *migrational effect,* these documented changes in the first generation offspring of migrants, to the improved environmental, mainly dietary, conditions of the New World? Moreover, the explanation could be extended to cover observed differences in succeeding generations of resident populations which underwent comparable dietary improvements through time. But newspaper accounts, historical records, even novels of the period, often suggest that the living conditions of migrants and their families–whether on the sugar plantations of Hawaii, in the railroad towns of the Western United States, or in the ethnic ghettos of the industrial Northeast–were often deplorable; the diet of migrants may have been no more adequate than that of their sedentary kinsfolk.

Since the documented differences between migrants and their offspring were well established, some investigators began to question whether migrants were really representative of the populations from which they derived. Was there something different, something special, about migrants which would help to account for the observed migrational effect? If mobile members of a population differ morphologically and genetically from the more sedentary individuals of the group, further differences could be expected between the offspring generations, even if environmental conditions remained similar for both groups. Different environmental settings would only accentuate these differences. This reasoning, applied in breed improvement practices, has been used successfully to develop distinctive lines of animal stocks.

Indeed, in 1954 Bernice Kaplan found morphological differences between migrants and the stationary population, "sedentes," from her analyses of a number of previous studies of migrant populations. Such differences, then, must be due to the distinctiveness of the migrant group and must represent a selected sampling from the gene pool of the ancestral group. Should interbreeding subsequently take place between migrants, even greater differences should appear between the offspring of migrants and the children of the sedentes. These differences could, of course, be further increased if improved environmental conditions were experienced as migrants' children were born and raised in the new countries. A hypothetical example may help to illustrate this process.

For simplicity's sake, let us assume that height, above a certain basic level of 155 cm., is influenced by two genetic loci. Each locus may have one of three allelic combinations present; A_1A_1, A_1A_2, or A_2A_2 and B_1B_1, B_1B_2 or B_2B_2. Height increments of 5 cm. result if the allele A_1 is present; of 4 cm. for the B_1 allele; of 2 cm. for the A_2 allele; and of 1 cm. for the B_2 allele. The possible allelic combinations and their total incremental values are indicated in Figure 6-1. If each genotype is represented in equal frequencies in this hypothetical population, the distribution of height in this population could be represented by the diagram in Figure 6-2.

A sample of migrants selected for higher incremental values, let us say of +18 cm. and +15 cm. (individuals whose stature is 170 cm. or

	A_1A_1 (10 cm)	A_1A_2 (7 cm)	A_2A_2 (4 cm)
B_1B_1 (8 cm)	$A_1A_1B_1B_1$ (18 cm)	$A_1A_2B_1B_1$ (15 cm)	$A_2A_2B_1B_1$ (12 cm)
B_1B_2 (5 cm)	$A_1A_1B_1B_2$ (15 cm)	$A_1A_2B_1B_2$ (12 cm)	$A_2A_2B_1B_2$ (9 cm)
B_2B_2 (2 cm)	$A_1A_1B_2B_2$ (12 cm)	$A_1A_2B_2B_2$ (9 cm)	$A_2A_2B_2B_2$ (6 cm)

A_1 = 5 cm
A_2 = 2 cm
B_1 = 4 cm
B_2 = 1 cm

FIGURE 6-1 Model of Height Increments, Polygenic 2-Locus System

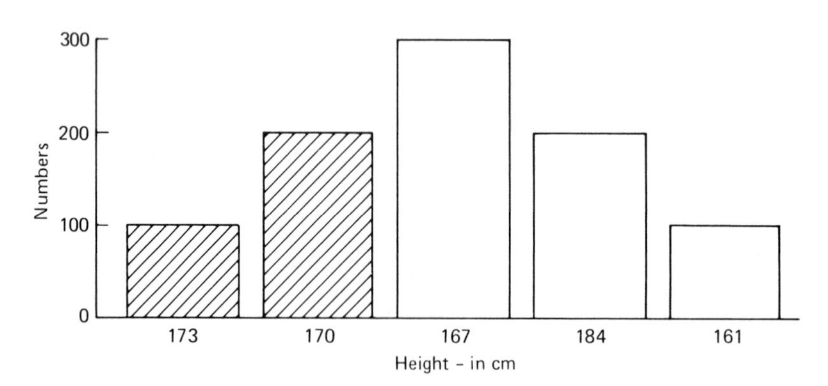

FIGURE 6-2 Model Diagram of Height Distribution

greater, as shown in the shaded portion of the diagram), would include individuals with the genotypes $A_1A_1B_1B_1$, $A_1A_2B_1B_1$, and $A_1A_1B_1B_2$, while the sedentary group would be represented by the remaining genotypes (the unshaded portion of the diagram). If these "tall" genotypes in the migrant population then cross only among themselves, the distribution of the resulting offspring generation will, ideally, be that shown in Figure 6-3. Thus, in this hypothetical migrant offspring generation, the shortest individual will measure 167 cm., and, if frequencies are calculated, about 45 percent of these individuals will be 173 cm. tall, 44 percent will be 170 cm., and only some 11 percent will belong to the shortest height category of 167 cm.

Many students of migrational effect have been concerned with *plasticity*, the ability to respond phenotypically to environmental pressures. We know, for example, that whatever the genotype for stature, a child who experiences chronic malnutrition or prolonged severe disease is unlikely to achieve phenotypic expression of the full genetic potential. Thus far, in our hypothetical case we have assumed that the phenotypes of the model directly reflected the genotypes found in the group. This simplifying assumption can now be examined by assuming that only a part, let us say one-half, of the incremental genetic potential was achieved by members of the ancestral model population in the homeland environment. Once again, the taller members of this group, those whose heights now exceeded 162 cm., migrated to another area and intermarried with one another. If this new environment allowed the expression of the full genetic complement of the offspring generation, the migrants' children (167 cm. to 173 cm.) would now exceed their parents (162.5 cm. to 164 cm.) in average stature and appear considerably taller, on the average, than their sedentary cousins.

		Gametes		
		A_1B_1 (9 cm)	A_1B_2 (6 cm)	A_2B_1 (6 cm)
Gametes	A_1B_1 (9 cm)	$A_1A_1B_1B_1$ (18 cm)	$A_1A_1B_1B_2$ (15 cm)	$A_1A_2B_1B_1$ (15 cm)
	A_1B_2 (6 cm)	$A_1A_1B_1B_2$ (15 cm)	$A_1A_1B_2B_2$ (12 cm)	$A_1A_2B_1B_2$ (12 cm)
	A_2B_1 (6 cm)	$A_1A_2B_1B_1$ (15 cm)	$A_1A_2B_1B_2$ (12 cm)	$A_2A_2B_1B_1$ (12 cm)

FIGURE 6-3 Genotypes, Migrant Offspring

Since the gene pool of the migrants' children now includes all four possible gametic combinations (A_1B_1, A_1B_2, A_2B_1 and A_2B_2), endogamous marriage in the migrant offspring generation will produce all the genotype combinations present in the ancestral population, although the frequencies of the genotype combinations contributing smaller incremental additions to total height will be relatively rare. For example, the $A_2A_2B_2B_2$ combination, with a total incremental value of 6 cm., can be expected to appear in only 2 of 1000 children (frequency = .002), and thus would be unlikely to appear in a small group of migrant grandchildren. It follows that the second generation of migrants' descendants will potentially include the entire range of variation for a metrical trait, but the average, or mean value, for that trait need not exceed the average of the migrant's first-generation offspring. In fact, Froehlich (1970) found that the growth trend for stature in male Japanese-Americans in Hawaii ended after one American-born generation; he ascribed the more gradual metrical increases in females, which extended into the third generation, to the greater cultural conservatism of females in the Japanese families.

In summary, it appears that the migrational effect involves both selective sampling of a migrant group and the effects of environmental factors on human plasticity. Recent studies suggest that new genetic combinations, as well as environmental changes, have also affected secular trends. Levels of inbreeding and consanguineous marriage have declined in many areas as technological advances have increased mobility and as family size has decreased. Migrants to cities have increasingly sought their marital partners in the metropolitan center rather than in the home village. Smaller family size has reduced the number of relatives from whom a spouse could be drawn. These factors have increasingly led to new kinds of matings and novel genetic combinations, as well as to exposure to varied patterns of behavior and environmental settings.

Heterosis, improved viability, growth or resistance due to genetic heterozygosity, is a phenomenon utilized by agricultural scientists to produce improved hybrid varieties of plants. Dahlberg suggested in 1943 that heterosis was an important component of the secular trend in stature among humans, and Hulse (1958), in his work on Swiss villagers, provided one source of evidence about heterotic changes in human stature (see Table 6-1). More recently, as noted earlier, Schreider (1967) demonstrated that there is a significant negative correlation between stature and the coefficient of inbreeding in seventy departments of France. In other words, adult stature, on the average, is lower for the offspring of genetically related parents than for the children of unrelated parents. Improved nutritional patterns and medical care, particularly as these may influence growth patterns, can contribute to the secular trend in stature, but genetic factors also play an important role.

TABLE 6-1 Body Size in Progeny of Endogamous and Exogamous Swiss Matings

Measurement	*Exogamous (Inter-village)*	*Endogamous (Intra-village)*
Stature (cm)	168.5	166.2
Weight (kg))	73.4	72.0
Shoulder breadth (cm)	38.8	38.7
Head length (mm)	189.0	187.5

From Garn, S. (1968), *Human Races,* 2nd ed., Springfield: C.C. Thomas, p. 106. Courtesy of Charles C Thomas, Publisher, Springfield, Illinois.

Just as the studies of migrational effects in humans have produced a wealth of published materials and opened up new areas of investigation, the study of intermixture has been richly productive and stimulating. Early discussions of race mixture were often accompanied by highly suspect conclusions which were scientifically unverifiable, as in Lord Kames's suggestion in 1774 that human racial diversity was the product of divine punishment. Daniel Brinton, an American anthropologist of the late nineteenth century, argued that the third generation of intermixture between whites and certain native groups (Polynesian, Australian, or Dravidian) were destined to extinction through short lifespan, feeble constitution, or sterility. Of course, Brinton and others who held similar views paid scant heed to the debilitating living conditions in which many of these offspring of mixed matings existed, and ignored the probable effects of such conditions on the human being of mixed or unmixed ancestry.

Scientific studies of intermixture, as opposed to armchair philosophizing or polemic discussions, began in the United States at least as early as 1894, when Boas published his work on "The Half Breed Indian." Stimulated by Professor Hooten's recounting of the genetics of the Rehobother Bastards, formed from unions of African Hottentots and the Boers of South Africa, Harry Shapiro began investigations on Pitcairn Island of the descendants of the English mutineers of the *Bounty* and their Tahitian wives; G.D. Williams initiated research on the Mayan-Spanish mestizos of Yucatan; Caroline Bond Day began her studies of the inheritance of skin pigmentation and hair form among American blacks in the United States.

Many of the issues which generated such heated discussions in the past as to the merits or disadvantages of intermixture can be seen, in the light of our modern understanding of genetics, to have often involved irrelevant, even spurious, questions. The immediate effect of intermixture between distinctive populations, each having achieved some degree of genetic adaptation, or fitness, in its own specific environment, is to increase genetic diversity and morphological variation. The hybrid population will receive alleles present in both parental groups, but these can now appear in novel genotypic combinations. Suppose that one parental population is homozygous for the *M* allele at the *MN* locus, while all members of the other parental groups have the genotype *NN*. Intermixture would be expected to produce offspring of all three genotypes: *MM, MN,* and *NN*. The new combination, *MN*, is now available for the operation of natural selection, and some evidence suggests that this genotype has a selective advantage in a number of populations. Again, it is clear that the heterozygote for sickle hemoglobin trait, genotype Hb_AHb_S, is at an advantage in areas where falciparum malaria is endemic, and mixture with a population in which the sickle allele is present would utlimately benefit a population homozygous for hemoglobin *A* where this form of malaria is prevalent.

Theoretically, it is also possible that intermixture might produce new genotype combinations which would be disadvantageous, and *Rh*-negative blood type has often been mentioned in this respect. Although the *Rh* blood group is rather complex, our purposes are served by considering this antigenic system to be controlled by three closely linked loci, with alleles at only one locus responsible for the most common *Rh* blood

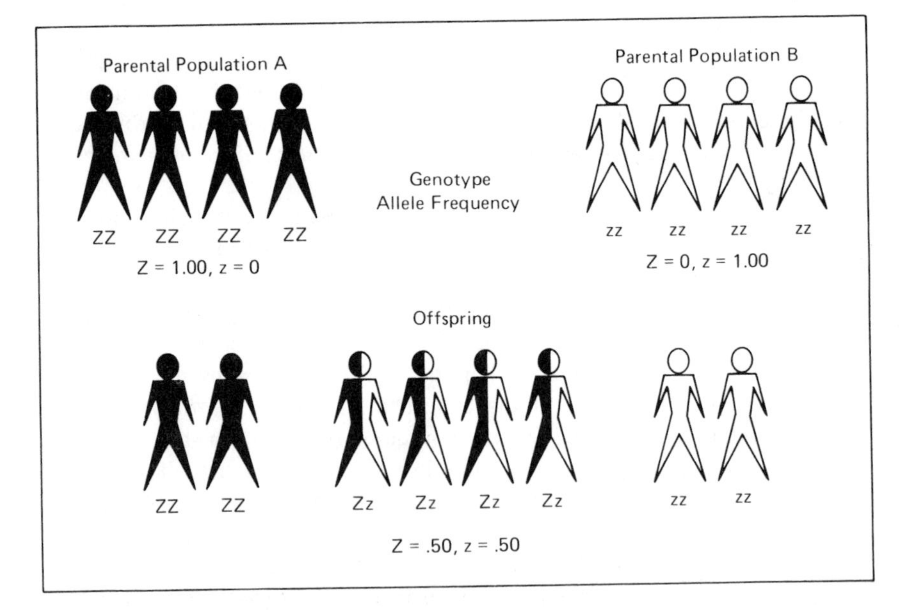

FIGURE 6-4 Diagram of Intermixture

incompatibility which may develop between a mother and her unborn child.

If a woman is *Rh*-negative, her genotype is *dd*, while that of her unborn child may be *Dd*, or *Rh*-positive. If blood from the fetus crosses the placental barrier and enters the mother's bloodstream, or if the mother has previously received *Rh*-negative blood either by blood transfusion or as the result of trauma during a prior delivery, her immunological system forms antibodies to the *Rh*-positive (*D*) antigen. If these anti-*D* antibodies cross the placental barrier and enter the bloodstream of the fetus, a pathological condition of the unborn child, erythroblastosis fetalis, may result. Today, children born with this condition can usually be successfully treated, and appropriate measures administered to *Rh*-negative mothers at the time of delivery can inhibit the development of anti-*D* antibody in the maternal blood. But only a few years ago, and even today in some parts of the world, maternal/fetal *Rh*-incompatibility has been subject to intense selective pressure.

The *Rh*-negative phenotype (genotype *dd*) is seldom, if ever, found in individuals of unmixed ancestry among natives of the Pacific islands, Australia, or in American Indians, although it is present in varying frequencies among European populations or those of European descent. Presumably, the *d* allele was rarely, if ever, present in the ancestors of these native peoples and, therefore, maternal/fetal *Rh* incompatibility was infrequent or unknown until after intermixture with Europeans occurred. Obviously, the incompatibility would also not occur in a population homozygous for the *d* allele; but no such group has ever been identified, although Boyd once postulated such a hypothetical population of archaic Europeans.

The incompatibility can only occur in a population in which both alleles are present, and the frequency with which the condition appears is a function of the relative frequencies of the two alleles, as Table 6–2 (which holds all other factors constant) indicates.

At low frequencies, recessive alleles are found predominantly in the heterozygous genotype combination, and the homozygous recessive trait occurs at very low frequencies. Thus, even when the *d* allele had a frequency of 0.20, only 4 percent of the population could be expected to express the *Rh*-negative blood type, and, in the diagram figures, the majority of the population should be *Rh*-positive even when the *d* allele frequency exceeds 0.50. These relationships hold true, of course, for many traits produced by a homozygous recessive genotype, but the implications are of immediate consequence in dealing with a condition which may affect so many women of Euroamerican or mixed descent and their children.

TABLE 6-2 Rh-incompatibility Risk as a Function of Allele Frequency and Mating Type Frequency for dd Females

Allele Frequencies		*Expected Genotype Frequencies* $(p^2 + 2pq + q^2)$			*Expected Mating Frequencies of dd* ♀ × *non-dd* ♂		
D	*d*	*DD*	*Dd*	*dd*	♀ *dd* × *DD* ♂	♀ *dd* × *Dd* ♂	Σ
.00	*0*	1.00	*0*	*0*	–	–	–
.80	.20	.64	.32	.04	.0256	.0128	.0384
.60	.40	.36	.48	.16	.0576	.0768	.1344
.40	.60	.16	.48	.36	.0576	.1728	.2304
.20	.80	.04	.32	.64	.0256	.2048	.2304
0	1.00	*0*	*0*	1.00	–	–	–

The risk of maternal/fetal incompatibility also depends on the frequency of matings between *Rh*-negative (*dd*) females and *Rh*-positive (*DD* or *Dd*) males. Assuming ideal conditions (complete panmixia, equal sex ratios, etc.), 4 percent of a population with allele frequencies of $D = 0.80$ and $d = 0.20$ can be expected to have *Rh*-negative blood type. Women with this genotype, married to men with the *DD* genotype, must necessarily have children with the *Dd*, or *Rh*-positive, type, and, therefore, all such pregnancies would be at risk for the maternal/fetal *Rh* incompatibility. Given the frequencies of these genotypes in this ideal population, about three (2.56) of every 100 marriages is likely to be such a "high risk" mating. Only about one (1.28) of every 100 marriages can be expected to occur between a *dd* female and a *Dd* male, and there is a probability of one-half that an offspring from the latter mating type will be *Rh*-positive (*Dd*).

When the allele frequencies were $D = 0.60$ and $d = 0.40$, about six of every 100 matings (.0576) involve the *dd* × *DD* mating combination which produces only *Rh*-positive children, but nearly 8 percent (.0768) of all matings expected would involve the pairing of *dd* females with *Dd* males, and the production of offspring of whom some may be of the compatible *dd* genotype. As Table 6-2 shows, the probability of all "risk" matings increases dramatically with the increased frequency of the recessive allele up to a certain level, but lower risk matings (♀dd × Dd♂) assume a greater proportion of all such matings as the recessive allele becomes more frequent. Several explanations have been suggested to account for the maintenance of heterozygosity at this locus in many human populations, and some of these will be discussed in a later chapter on natural selection. For now, it is useful to note that when the *d* allele is present at a low frequency, relatively few incompatible matings are likely to occur; most of these will involve the mating of *dd* women with *DD* men, and, thus, all such fertile matings will produce fetuses at risk for maternal/fetal *Rh* incompatibility.

Intermixture between an Indian population lacking the *d* allele and any one of a number of Euroamerican groups in which the allele is present would produce an offspring generation in which the allele *d* is present at low frequencies. Suppose that a population with allele frequencies of $D = 1.00$ and $d = 0.0$ mixes with a numerically equal group in which the *d* allele has the value of 0.20. The new gene pool ($D = 0.90$, $d = 0.10$) would produce, with random mating and under ideal conditions, offspring with the genotype distributions $DD = 0.81$, $Dd = 0.18$, and $dd = 0.01$. No part-Indian would have the *dd* genotype in the first offspring generation since these cannot appear in any cross of a *DD* Indian with non-Indian (*DD, Dd,* or *dd*) mates. Subsequent generations could include females with the homozygous recessive condition, but because of the low frequency of the *d* allele, matings of these rare *dd* women will usually involve males of the *DD* genotype. Children from such matings are, as we have already seen, necessarily *Rh*-positive, and at highest risk for the operation of natural selection. It follows, then, that this oft-cited case of the "dangers of intermixture" is not so simple as it might at first appear. The first descendant generation will have a lower frequency of *dd* individuals than were present in the Euroamerican parental group, and no part-Indian *dd* individuals will occur. Even in subsequent descendant generations, *dd* females will be rare and the proportion of their Indian ancestry increasingly less.

Historically, mixture has often been less complete than this illustration suggests, and it has involved the contributions of relatively few individuals, most often males, to the gene pool of another population. Such men, even if they are of the rarer *dd* genotype, can only produce *Dd* children by native wives from groups homozygous for *D*. At the low allele frequencies which would result from this kind of admixture, very few, if any, individuals with the *dd* genotype would appear in later generations unless a pattern of assortative mating between persons with mixed ancestry developed. In the latter case, the *dd* genotype could be expected to occur more frequently than in the rest of the population and would be subject to removal through the operation of natural selection, especially in the absence of advanced medical knowledge and modern practices. This may help to account for the seeming paradox of the rare frequency of *Rh*-negative donors found by Myrianthopoulos and Pieper at the Guam Memorial Hospital Blood Bank.

Guam is the largest of the Mariana Islands in the northwestern Pacific, with a total land area of some 215 square miles. The island was discovered by Magellan in 1521, but a permanent Spanish colony was established there only in 1668. In the following years the native population, numbering at least 30,000 when discovered, was decimated, but Spanish, Mexican, Filipino, and Carolinian migrants, as well as a few Europeans and Americans, visited or settled on the island throughout the period of Spanish control, which ended in 1898. When these blood typing studies were done, only two of 1497 donors were found to be *Rh*-negative. Myrianthopoulos and Pieper, noting that the frequency of the *Rh*-negative allele among modern Spaniards ranges from 30 to 40 percent, were led by their findings to question the then-current notion that Spanish contributions to the native, or Chamorro, gene pool had been extensive; later studies confirmed their suspicions (Underwood, 1973, 1976). Today, only one of every 1000 Chamorros can be expected to have the *dd* genotype, or thirty-five out of a 1960 census population of 34,762 Chamorros, of which only 17 or 18, predictably, would be females with the homozygous recessive genotype. There is, however, some evidence that preferential marriage patterns formerly existed between families with Spanish ancestry, so that natural selection has undoubtedly operated against the *d* allele and its present frequency is somewhat misleading in calculating the degree of Spanish admixture.

Finally, and of great importance in considering the consequences of intermixture, there is the role of genetic variation in a long-range evolutionary perspective. The evolution of a number of populations, each adapted to a specific ecological niche, implies a possible loss of genetic variation within each group, possibly for the species as a whole. At a moment in time, in the particular environment, each population may be said to have achieved an adaptive peak. But environments change and the alleles formerly lost through natural selection or by fixation as a result of genetic drift may be the very ones which are responsible for a phenotype which would prove selectively advantageous under these altered conditions.

The retention of the allele for sickle hemoglobin at low frequencies in a population not exposed to falciparum malaria provides a hedge against the demise of the group, should this form of malaria become endemic in their area. The cost seems heavy for this kind of protection against conditions which may never be experienced, but from an evolutionary perspective, the alternative is even more disastrous. Intermixture, which results in increased genetic variability, operates to make new genotypic combinations available which may offer advantages not experienced by the parental populations, and helps to maintain variation in the species as a whole.

Now, having already considered some of these issues in the study of intermixture, the complex genetic composition of the American Black can be approached more readily. Although a few of the African ancestors of American Blacks came to the United States before 1700, most slaves were brought into this country in the period between 1700 and 1808, at which latter date importation became illegal, and over 98 percent came from West and west-central Africa. In the absence of detailed historical accounts of the nature and extent of interracial matings during most of this period, many attempts have been made to calculate the degree of admixture of non-African genes into the gene pool of the American Black. In theory, the calculation is fairly straightforward, since the amount of mixture (M) is estimated by dividing: (1) the allele frequency at a particular locus in American Blacks (q_{AB}) less the allele frequency for the same locus in a representative African population (q_{AF}); by (2) the allele frequency of this gene in the non-black ancestral population (q_{AW}) less the allele frequency at that locus in the representative African population (q_{AF}):

$$M = \frac{(q_{AB} - q_{AF})}{(q_{AW} - q_{AF})}$$

Assume, for a certain locus, that American blacks have an allele frequency of 0.50, a modern West African population a frequency of 0.0, and a modern American white population has an allele frequency of 1.00. Substituting in the formula:

$$M = \frac{(0.50 - 0.0)}{(1.00 - 0.0)} = \frac{0.50}{1.00} = 0.50.$$

Our estimate would show an admixture value of 0.50, and we might, in foolish haste, conclude that 50 percent of the American Black gene pool was derived from American Whites. We might even compound this folly and estimate the average rate of admixture per generation. If we allow twenty-five years per generation, roughly ten generations have elapsed between 1700 and 1950 (250/10 = 25), so the average rate of admixture would have been 5 percent per generation (M = 0.50/10 generations). And some earlier investigators concluded from their calculations that the admixture values of 20 to 30 percent in some tested American Black populations represented the actual value of American White contributions to the gene pool of American Blacks.

As Reed (1969) has pointed out, the criteria for applying this formula were not always recognized or met. First of all, the composition of the two ancestral populations was not always known. In 1949, Meier found that a large percentage of American Black college students claimed Indian ancestry. A few years later, Glass (1955) provided evidence to contradict this claim, and we still know little of the detailed ancestry of the Africans from the various slaving areas, or of the detailed ancestry of the Europeans who migrated to the New World and their descendants. How many native countries are represented in your own genealogy?

Second, since we cannot blood type our long-dead ancestors, it is necessary to use blood type frequencies of modern populations in West and west-central Africa and of American Whites in the United States in calculating admixture rates. This practice assumes a constancy which implies that evolutionary forces (selection, mutation, drift) have not been operating. As we shall see in a later chapter, selective pressures have probably affected the frequencies for all the major antigenic systems used in such calculations.

Until very recently, it was thought that the Duffy blood group locus provided an ideal trait for admixture studies in American Blacks. The Fy^a allele frequency for modern Africans in the slave area was less than 0.03, but, among American Whites, the frequency was about 0.43. Studies among California Blacks revealed no evidence of selection (decreased fertility or increased mortality), so the Duffy blood group allele frequencies were used to derive revised *M* values, supposedly devoid of the problems of natural selection. Although the *M* value of these reports varies somewhat from *M* figures obtained in studies using *ABO, Rh,* and other blood group systems, the marked difference between southern Blacks and northern Blacks remains. Recent evidence (Gelpi and King, 1976) suggesting that the Duffy blood group system is subject to selective pressures involving resistance to vivax malaria and, perhaps, falciparum malaria, challenges the exact values for *M* in these studies, but does not invalidate these patterns of differences. American Blacks are a national minority, composed of a number of breeding populations which vary from one to another in allele frequencies and in the degree of admixture represented in each group's gene pool. Some Black groups show little admixture, and this may well reflect the genetic history of such groups; others may have lost, through migration or through movement into the White population, individuals with a greater amount of White ancestry.

What, then, can we conclude from the host of studies dealing with the genetic composition of American Blacks? Much remains to be done in reconstructing the historical details of the process of admixture which has undeniably taken place over the last few hundred years, and genetic studies have a contribution to make to this research. One recent study of X-linked trait frequencies has supported historical evidence that American White females contributed to the American Black gene pool (Dyer, 1976).

Our increasing knowledge of the kinds and rates of selection intensities at various genetic loci may help us to make more sophisticated estimates of the degree of admixture which has already occurred—and it *has* been a two-way process. We have barely begun to look at admixture from the American Black gene pool into the gene pool of American Whites, but we definitely *share* genes; we are indeed all brothers and sisters. Everything we know from a theoretical standpoint enables us to say that this exchange has increased genetic variation in both groups and has provided both with enhanced evolutionary potential.

PROBLEM SET: GENE FLOW

---

1. An itinerant group of Basque sheepherders settled among a group of Native American Indians and, over the years, the sheepherders took wives from among the Indian women. The original group of Basques included 22 men with the *Rh*-negative phenotype, while the remaining 23 Basque men had *Rh*-positive blood type. Since all the foreign men married one Indian woman each, the gene pool of this group of Basque men and Indian women, a total population of 90 adults, could be expected to have the approximate allele frequencies of D = ________ and d = ________.
2. Assuming that the parental population described above mated and each couple produced an equal number of children, what are the expected phenotype frequencies of the F_1 generation?
   Rh positive ________ *Rh* negative ________.
3. If the Indian-Basque children of the F_1 generation all survive to adulthood and random mating prevails, an equal number of children being produced by each couple, the phenotype frequencies for *Rh* blood type can be expected to approximate the following values:
   Rh positive ________ *Rh* negative ________.

REFERENCES AND RECOMMENDED READINGS

BOAS, F. 1894. The half blood Indian. *Popular Science Monthly* 14:761.

——. 1911. *Changes in Bodily Form of Descendants of Immigrants.* Washington: Government Printing Office.

DAY, C.B. 1932. *A Study of Some Negro-White Families in the United States.* Cambridge, Mass.: Peabody Museum.

DUNN, L.C. 1959. *Heredity and Evolution in Human Populations.* Cambridge, Mass.: Harvard University Press.

DYER, K.F. 1976. Patterns of gene flow between Negroes and Whites in the United States. *Journal of Biosocial Science* 8:309–333.

FRISCH, R., and R. REVELLE 1971. The height and weight of girls and boys at the time of initiation of the adolescent growth spurt in height and weight and the relationship to menarche. *Human Biology* 43:140–159.

FROEHLICH, J.W. 1970. Migration and the plasticity of physique in the Japanese-Americans of Hawaii. *American Journal of Physical Anthropology* 32:429–442.

GELPI, A.P., and M.C. KING 1976. Duffy blood group and malaria. *Science* 191:1284.

GLASS, B. 1955. On the unlikelihood of significant admixture of genes from the North American Indians in the present composition of the Negro of the United States. *American Journal of Human Genetics* 7:368–385.

HULSE, F. 1958. Exogamie et heterosis. *Archives Suisses d'Anthro. Gen.* 22:103–125.

KAPLAN, B. 1954. Environment and human plasticity. *American Anthropologist* 56:780–799.

MEIER, A. 1949. A study of the racial ancestry of the Mississippi college Negro. *American Journal of Physical Anthropology* 7:227–240.

MYRIANTHOPOULOS, N., and S.J.L. Pieper, Jr. 1959. The ABO and Rh blood groups among the Chamorros of Guam, with reference to anthropologic and genetic problems in the area. *American Journal of Physical Anthropology* 17:105–108.

REED, T.E. 1969. Caucasian genes in American Negroes. *Science* 165: 762–768.

SCHREIDER, E. 1967. Body-height and inbreeding in France. *American Journal of Physical Anthropology* 26:1–4.

SHAPIRO, H.L. 1936. *Heritage of the Bounty*. New York. Doubleday and Company.

UNDERWOOD, J.H. 1973. Population history of Guam: context of human microevolution. *Micronesica* 9:11–44.

——. 1976. The native origins of the neo-Chamorros of the Mariana Islands. *Micronesica* 12:203–209.

WILLIAMS, G.D. 1931. *Maya-Spanish Crosses in Yucatan*. Cambridge, Mass.: Peabody Museum.

CHAPTER 7

Mutations, Radiation, and Human Variation

To those of us who are science fiction buffs, the mutant life form—a variant whose distinctiveness from familiar organisms is a function of the imagination and creativity of the skilled writer—is a commonplace feature of the genre. Many novels of this kind begin with or shortly after a cataclysmic event, perhaps a nuclear holocaust so serious as to cause the massive genetic changes required to thus drastically alter the ongoing life processes with which we are familiar. But scenes of this kind, and the creatures invented to inhabit a planet which has undergone such a catastrophe, bear little resemblance to all we know about ongoing mutation or, so far as we can judge from the fossil record, to the operation of this evolutionary force in the history of the human species.

In the broadest sense, mutation consists of changes in the genetic material. To be quite honest about it, what we know about mutation concerns only those genetic changes which produce phenotypic alterations or differences which we can somehow detect. There may well be genetic changes which result in no detectable differences in the phenotype, and these remain simply unknown to us, at least until such time as we develop methods capable of measuring subtle differences in the molecular composition and structure of the genetic material. For the time being, however, we can recognize several forms of genetic changes with observable phenotype effects. These include a range of modifications in the genetic material, from the level of the substitution of a single base in a nucleotide of nucleic acid (*point mutation*) to that of *polyploidy,* a multiplication of a whole chromosomal set. Only in a very general way do the resultant phenotypic changes parallel the extent of the genetic modifications. A base substitution, changing a single amino acid in the lengthy sequence of the components of a polypeptide chain may be as lethal to the development of a viable human organism as would be the presence of a triploid (3N, or 69) chromosomal complement in the body cells. But in general, at least in most animal species, minor changes, point mutations, are more likely to produce slight phenotypic changes which permit the survival of the mutant organism than are gross alterations in the chromosomal complement which very often seem to result in imbalances affecting the entire developmental pattern. Improved breeds of certain ornamental plants, the result of tetraploidy, have had commercial value, but viable mutant forms of polyploid humans seem to be confined to the pages of fictional novels.

Mutations may be produced by a number of factors, *mutagenic agents,* including some chemicals and certain forms of radiation. In addition, it now seems likely that certain viruses may also act as mutagens, either by inducing biochemical changes in the DNA of a host organism or by inserting segments of their own DNA into that of the host. Changes of the genetic material which affect the somatic cells may have significant effects on the functioning and well-being of the individual, but only those mutations which affect germ cells have evolutionary value, since these alone can be passed on to the gene pool of succeeding generations. Because the gonads, particularly the ovaries of the female, are relatively well protected within the body, germ cell mutations are less frequent than those found in cells in the more exposed parts of the body. A standard dose of X-ray irradiation in the course of dental diagnosis, for example, results in a negligible dose to the gonads if proper precautions are taken.

Mutations can be experimentally induced by the application of known mutagenic agents, but some genetic changes may occur as a result of exposure to natural radiation from cosmic rays, radioactive materials present on the surface of the earth, and radioactive elements present in some of the foods we eat. These have often been referred to as spontaneous mutations. Ionizing radiation, that is, radiation capable of producing changes in the electrical charge of atoms, includes the effects of X rays or gamma rays. Mutations can also be produced by non-ionizing radiation sources, such as ultraviolet rays, the visible light spectrum, and infrared rays.

One of the measures of radiation exposure is the *roentgen (r),* the amount of radiation required to produce 2×10^9 ion pairs per cubic centimeter of air. Some estimates suggest that we are exposed, on the average, to about 3 or 4 roentgens every generation, or roughly 10 r. over a seventy-five year lifespan, from natural radiation. Under ordinary circumstances, each of us may also receive a total dose of 25-30 roentgens over the same time period in the course of medical diagnostic and treatment procedures. Obviously, these values change if extensive medical procedures involving radiation are used during the period or if the individual is engaged in an occupation involving increased risks of exposure to radiation sources. Moreover, geologic variations cause differences in the amount of radiation present in the earth's crust in various parts of the world, while the amount of cosmic rays increases with altitude. Accordingly, an estimate of the average exposure received in a lifetime is intended to serve only as a crude approximation. Using these figures, we might estimate that 30-50 r. represents a hypothetical lifetime risk value for radiation exposure from known major sources.

Doses as low as 10 r. may cause pathological changes in the *leukocytes,* or white blood cells, of humans, and over 50 percent of the adults of Rongelap Island who were accidentally exposed to over 200 *rads*[1] from nuclear test fallout were found to have chromosomal abnormalities. The incidence of leukemia has been much higher in survivors of the atomic bombing of Hiroshima than in the Japanese population as a whole. But all these cases have involved an intense, acute dose of radiation received in a single administration. Experimental studies with small mammals indicate that a given amount of radiation administered over a period of time (chronic dose) results in fewer mutations than if administered in one brief exposure ("acute dose").

Whatever the mutagenic agent, the effect of a mutation in a germ cell line is to introduce new genetic material into the gene pool. Mutations, then, are the ultimate source of genetic variation. In recent decades, however, public concern has been directed toward limiting needless exposure to mutagenic sources of all kinds and, thus, reducing the incidence of new mutations. There are two explanations for this concern—one having to do with the observed phenotypic effects of most recognized mutant alleles, and the other based on a growing recognition of the vast amount of genetic variability already existing in human populations.

So far as we know, all genes are subject to mutation, and mutation is itself a random event in the sense that any locus is at risk for change

[1] A rad is a unit of absorbed radiation, roughly equivalent to the amount of ionization expected from a dose of 1 r.

as a result of exposure to a mutagenic agent. In the course of the several million years of hominid evolution, mutations have probably occurred at a rate at least no greater than can be observed in the present. But what happens to mutations after they are incorporated into the gene pool is not random. Natural selection, operating on the phenotype produced by the new genotype combination, determines the fate of the mutant allele in a predictable direction. Presumably, mutations which have conferred a selective advantage in the course of human evolution have already been incorporated into the gene pool of the species by means of natural selection. Conversely, mutations resulting in a selective disadvantage would presumably have been repeatedly removed from the gene pool.

The process proceeds most efficiently in respect to mutations for alleles with a pattern of dominant inheritance. A mutant dominant allele will be expressed in the phenotype whether present in homozygous or heterozygous combination, and is immediately subject to natural selection. If this allele produces a *lethal* condition, one which results in the death of the organism before or shortly after birth, the allele will be removed soon after it is introduced into the gene pool.

A mutation resulting in an allele whose products are only expressed in the homozygous recessive combination is not subject to the operation of natural selection until it appears in the company of another recessive allele at that locus. The frequency with which this combination will occur is a function of the frequency of the allele in the gene pool as a whole. If the allele frequencies at a certain locus are $A = .90$ and $a = .10$, the homozygous recessive combination will, expectedly, appear in only 1 percent ($q^2 = .10^2 = .01$) of the population, and the recessive allele will most commonly be found in the heterozygote combination ($2pq = .18$) where it is not subject to selection.

If this line of reasoning is valid, the majority of mutations now appearing are unlikely to have selectively advantageous effects. On the contrary, new mutations would be far more likely to have deleterious effects on the overall adaptive fitness of living populations. This viewpoint has been endorsed by many notable scientists who point to the known genetic load of alleles with deleterious effects already present in human populations. It has also been argued that medical advances are even contributing further to that load by extending the lifespans of hemophiliacs, diabetics, victims of Down's syndrome, and others suffering from genetic diseases, to an age at which reproduction by the afflicted is possible.

The viewpoint bespeaks the humanitarian concerns of its supporters, but it also implies a static view of human evolution. Mutations which have a deleterious effect under one set of environmental conditions may provide for the survival of the group under different circumstances. Before the discovery of penicillin, most bacteria probably carried the genotype for sensitivity to penicillin (*PS*), but a few rare mutants must have occasionally appeared with the genotype for resistance to penicillin (*PR*). With the advent of widespread medical use of antibiotics, bacteria with the *PR* genotype survived while the *PS* variety succumbed to penicillin administration. Today, penicillin-resistant forms of bacteria have become prevalent among some human groups while penicillin-sensitive bacteria appear as rare mutant forms in these microbial populations. A somewhat parallel sequence of events probably characterized the spread of the sickle hemoglobin allele in certain African tribes as they moved from an economy based on hunting and gathering to the pursuit of horticultural activities.

The same objection may be raised in respect to the argument that genetic variability is already so great that recombinations of all existing allele combinations can provide all the genetic variation required for the survival of the species under any conceivable set of environmental circumstances. At the gross level of chromosomes, millions (2^{23}) of unique combinations of maternal and paternal chromosomes are possible. The number of possible novel gene combinations is even greater, of course; so adherents of this view argue that the genetic variability of our species has barely begun to appear in the few billion members of our species who have ever lived. This is undoubtedly true, but it is unreasonable to expect that we can so surely predict environmental circumstances in a few million years from now as to be certain that new mutations cannot possibly prove essential to the survival of the species at that distant period when we may even be inhabiting other planets. The mutational load of today and tomorrow exacts a heavy price in personal suffering and tragedy, as well as financial expense, about which any humane person is concerned. But the alternative, the possible extinction of the species, should not be ignored in the face of immediate costs. The needless induction of new mutations is surely unwarranted, but it is both impossible and lacking in an evolutionary perspective to try to ban the use of any mutagenic agent for any purpose.

In any case, mutations must be relatively rare events, since, at a given time, any random modification of an intricate functioning system evolved over countless generations is more likely to prove dysfunctional than beneficial. The exact rate of mutations in humans may never be known because of the difficulties; of detecting newly mutant recessive alleles when present in a heterozygote genotype; of distinguishing genetic changes from somatic changes which affect the embryo at an early stage of development; of identifying dissimilar genotypes producing similar phenotypic effects (genic heterogeneity); and in a number of other conditions which obscure specific identification of mutant alleles. However, estimates of mutation rates (Table 7–1) do suggest a range of likely gamete mutation rates, ranging from five to perhaps 100 per million gametes each

TABLE 7–1 Estimates of Human Mutation Rates*

Trait	*Mutations per Million Gametes per Generation*	*Estimated Fitness*
Retinoblastoma Dominant; an eye tumor	15 - 23	0
Juvenile amaurotic idiocy Recessive; blindness, paralysis, mental deficiency, death, onset at about 6 years of age; common in Scandinavia	38	0
Infantile amaurotic idiocy (Tay-Sachs disease) Recessive; symptoms as above, but onset around 2 years of age; common in Jews	11	0
Microcephaly Recessive; abnormally small skull	49	~0
Achondroplastic dwarfism Dominant	10 - 70	0.1
Hemophilia Sex-linked	25 - 32	0.25 - 0.33
Muscular dystrophy Sex-linked	43 - 100	0.30
Albinism Recessive	28	< 1.0
Aniridia Dominant; absence of iris	5	?
Deaf-mutism Several loci	450	–
Low-grade mental defect Many loci	1500	–
All loci causing death before early adulthood	40,000	0

*The mutation rates given here for dominant and for sex-linked genes are much more accurate than those for recessives. Many human geneticists view estimates for recessives with great skepticism, because of the uncertainties inherent in current methods of determination. They are given here merely to indicate the magnitude of the estimates made. (From Lerner & Libby, 2nd ed. 1976.)

generation. If the number of genetic loci in humans is somewhere between 100,000 and 1,000,000, and each individual is formed from the fusion of two gametes, there is an appreciable chance that each of us carries several newly mutated genes.

Most, probably all, mutations are recurrent, and the change from a dominent allele to a recessive allele, for example, occurs repeatedly through time. Ignoring, for the moment, any possible selective advantage or disadvantage, it is possible to calculate the number of generations which would be required in order for recurrent one-way mutation (A→a) to replace the dominant allele by the recessive allele. For our purposes, this can be represented by considering a population to be homozygous for the *A* allele (*A* = 1.00) and to experience an unrealistically high mutation rate (u) of 10 percent. In one generation, the change in the frequency of the *A* allele (Δp) will be equal to minus u times p, or: $\Delta p = -up$. This can be shown for a few generations, as in Table 7-2. It should be noted that as fewer *A* alleles are available to be mutated from *A* to *a*, the rate of change declines, as shown in Figure 7-1. Actually, the tedious calculation of change in allele frequency from one generation to the next can be avoided by the use of a simple formula, $p_n = p_o\,(1-u)^n$, where n = the number of generations. In order to calculate the rate of change for the fifth generation, the numbers presented earlier are merely substituted in this formula:

$$p_5 = 1.00\,(1 - .10)^5 = 1.00\,(.90)^5 = 1.00\,(.059049) = 0.59.$$

If recurrent, one-way mutation continued over a sufficient number of generations, changing *A* to *a*, the population would ultimately become homzygous for the *a* allele.

Of course, it is wholly unrealistic to assume that mutation is entirely unidirectional. Rather, as some alleles are mutating from *A*→*a*, reverse

TABLE 7-2 Change in p Due to One-Way Recurrent Mutation, u = .10

Generation	p_A	q_a
p_0 (parental)	1.00	0.0
p_1	0.90	0.10
p_2	0.81	0.19
p_3	0.73	0.27
p_4	0.66	0.34
p_5	0.59	0.41

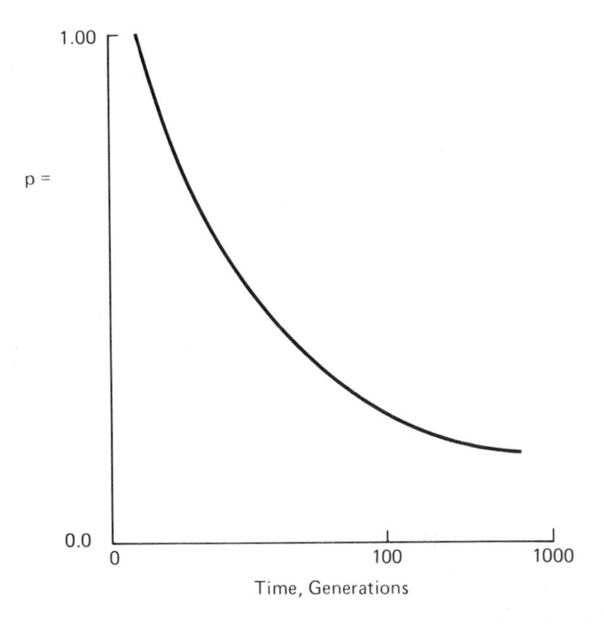

FIGURE 7–1 Change in Allele Frequencies as a Result of Recurrent Mutation $A \rightarrow a$, where $u = .10$

mutation, $a \rightarrow A$, is also taking place. Again, for purposes of illustration, consider a model of this process in which the initial values of the two alleles at a locus are $p = 0.50$ and $q = 0.50$, and A alleles are mutating to a alleles at a rate of 10 percent per generation ($u = 0.10$) while a alleles are changing to A alleles at the rate of 5 percent per generation ($v = 0.05$). Then the change in the frequency of the A allele each generation is expressed by the formula

$$\Delta p = -up + vq.$$

A simplified chart covering several generations of recurrent two-way mutations at these rates can illustrate the major predictable trends of such changes (see Table 7–3).

If the Table 7–3 model were carried through enough generations, and the recurrent reversible mutation rates remained unchanged, an equilibrium point would be reached when $p = 0.33$ and $q = 0.67$. At this point, determined solely by the value of the two mutation rates, a genetic equilibrium would have been attained and this equilibrium would remain stable so long as the mutation rates were unchanged. In other words, whatever the original values of the two alleles, equilibrium is reached

TABLE 7-3 Change in p Due to Recurrent Reversible Mutation (u = 0.10, v = 0.05)

Generation	p_A	q_a
0 (parental)	0.50	0.50
1	0.475	0.525
2	0.454	0.546
3	0.436	0.564
4	0.421	0.579
5	0.408	0.592

and maintained at a point determined exclusively by the value of the two mutation rates. This is somewhat akin to the problems of maintaining minimal balances in both a checking (p) and a savings (q) account when one regularly moves, let us say, 20 percent of the monthly balance (u) from savings to checking and 10 percent of the monthly balance (v) from checking to savings account, as shown in Table 7–4.

As we have already seen, the fate of mutations in human populations is not merely a function of the frequency of their occurrence, but depends on the selective advantage or disadvantage which the mutant phenotype has. Models such as these assume that no other evolutionary force is operating, but the equilibrium point actually attained will depend also on the value of s, the coefficient of selection. The relationship between s and the mutation rates u and v is more fully discussed in a later chapter. We can turn for the remainder of this chapter to a review of some of the studies which have been made on the role of mutation in the maintenance of variation in human populations.

Obviously, we cannot intentionally conduct mutation experiments on humans, so much of our information on the subject has come from studies in groups exposed to unusual occupational risks, in populations living in areas with high levels of radiation exposure, among individuals who have undergone certain kinds of medical treatment, or in those who have been accidentally subjected to high levels of exposure to mutagenic agents. Among groups subject to unusual occupational risks, it has long been known that medical radiologists have a higher incidence of leukemia than is found in the general population or in an unexposed control group matched as to age, sex, or economic status. Similarly, the high incidence of cancer in women formerly employed to place radium paint on watch dials has been attributed to the ionizing effects of this radioactive material. It might also be noted here that Marie Curie, who

TABLE 7-4 Hypothetical Bank Accounts, a Wild Model of Recurrent Reversible Exchange Between Savings and Checkings Accounts. (u = .20, v = .10)

Month (= generation)	*Checking Account (= p)*		*Savings Account (= q)*	
0	$70		$30	
		+3		+14
		-14		-3
1	59		41	
		+4		+12
		-12		-4
2	51		49	
		+5		+10
		-10		-5
3	46		54	
		+5		+9
		-9		-5
4	42		58	
		+6		+8
		-8		-6
5	40		60	
		+6		+8
		-8		-6
6	38		62	
		+6		+8
		-8		-6
7	36		64	
		+6		+7
		-7		-6
8	35		65	
		+7		+7
		-7		-7

with her husband discovered radiation, and her daughter, Irene, died of radiation-induced disease.

In addition to occupation risks, it has also been suggested that populations living in areas with relatively high levels of background, or natural, radiation from the rocks and soils in the local region could be expected to show a higher incidence of mutations than would groups living in areas where the radioactive content of the physical setting is lower. Some areas of upper New York State have such geological formations rich in rocks containing radioactive elements. Guthrie and his co-workers (1959) found a marked association of increased congenital malformation rates with residence in areas with relatively higher levels of radioactivity, while no comparable statistical association was found between congenital malformation rates and other possible causative factors such as medical radiation exposure, socioeconomic status, rubella measles incidence or rural/urban differences. The authors of the study suggested that radiation is the most likely causative agent of increased congenital information rates in New York State.

By way of contrast, studies have been made of the wild rat population of Kerala province in India. In this area, extremely high levels of background radiation characterize the sands which are rich in the radioactive element thorium. When compared to wild rats in neighboring areas, no significant phenotypic differences were found, although the local rat population must have been exposed to radiation levels several times higher than those experienced by their more fortunately situated neighbors. While it is always unwise to extrapolate directly to humans from studies in other animals, these findings suggest that chronic, low-level radiation dosage over many generations may result in mutation rates unlike those found in populations which experience high level, intense radiation exposure. Indeed, a comprehensive study of variation in neonatal death rates (an available indirect measure of radiation effects) in the Western United States uncovered a relationship between deaths of newborns and altitude. But Grahn and Kratchman (1963) argue that this is not due to higher levels of terrestrial radiation; instead, they argue that the significant correlation is between birth weight and altitude, and that low birth weight at higher altitudes can be related to decreased oxygen partial pressure. These authors conclude that reduced oxygen partial pressure at higher elevations results in reduced fetal growth, and that low weight or underdeveloped fetuses are at greater risk at and immediately after birth than are larger and more fully developed newborns.

In addition to radiation effects, a number of studies have found evidence of chromosomal changes in body cells induced by exposure to various chemicals. Claims have been made, but also challenged, that such drugs as LSD or the active element of marijuana, THC, have this effect. But a careful ongoing series of studies of several human populations exposed to methyl mercury through fish consumption has confirmed the correlation between the frequency of body cells with chromosomal breakage and mercury concentration of the red blood cells. Many other chemicals have been claimed to have a mutagenic effect in experimental animals, but the doses used in these experiments have often been higher than the levels to which humans are ordinarily exposed, and most studies have been conducted on small laboratory animals which differ from us in size, anatomy, and also in some significant physiological traits. Some somatic mutations may well be caused by chemicals we ingest, but little is yet known about the exact amounts, necessary length of exposure, or specific metabolic pathways involved in humans, to determine safe limits of exposure. Because of the possible hazards to the developing embryo and fetus in the critical early period of intrauterine life, it is probably wise for every expectant mother to avoid any suspect substance during at least the first trimester of pregnancy.

Ample evidence exists that exposure to X-ray irradiation is capable of causing mutations in humans, and medical treatments involving X rays usually involve acute exposures of at least several roentgens per minute. Stern (1973), by making several simplifying assumptions, estimates that an irradiation dose of 10 r. at high intensity to both parents of 1000 children would result in four of those children carrying an induced mutation, although most of these mutations probably will not be visibly expressed in the offspring. Because of the difficulties of detecting specific phenotypic effects of mutations, especially those involving changes to a recessive allele, most studies have relied on more general evidence of radiation effects, such as increased mortality rates in subjects exposed to medical sources of radiation or in their offspring. For example, Meyer and his associates examined the effects of X-ray exposure on women who had been exposed *in utero,* before birth, and found that these women, as adults, had a significantly higher number of sons than daughters, longer gestation periods for their offspring, lower placental weights, and more precipitate labors and deliveries when giving birth to their own children.

Several human populations have been exposed to high intensity, acute radiation as a result of atomic bomb explosions. Extensive early studies of the survivors of the Hiroshima and Nagasaki bombings in 1945 provided ambiguous evidence on the gametic effects of intensive radiation exposure. The percentage of congenital abnormalities in children born to couples of whom one or both had been exposed was even slightly less than the percentage of congenital malformations born to couples, neither of whom had been exposed to radiation from the atomic bombings. By 1953, no significant increase in rates of malformation, stillbirths, or deaths of newborns was found in children conceived after one or both of the parents had been exposed to radiation. However, continuing studies have clearly demonstrated a number of somatic effects. Among those who were less than thirty-one years old at the time and received a dose of about 200 r. or more, 34 percent showed chromosomal changes, while the percentage of those older than thirty at the time of exposure who exhibited chromosomal abnormalities was even higher. Additionally, statistically significant decreases in body measurements and post-pubertal growth rates of exposed children were found in 1953. Most strikingly, acute leukemia rates and thyroid cancer rates are higher for all age groups subject to ionizing radiation from the bombing explosions.

One attempt to find an indirect measure of radiation effects concerns the predictable effects of radiation on the sex chromosomes and *X*-linked genes. An exposed mother, whose sex chromosomes (as well as all her autosomal chromosomes) would have been subject to mutation damage, would transmit some affected *X* chromosomes to her offspring, male or female. Her irradiated *X* chromosome would be expected to have more serious consequences in her sons than in her daughters, since the latter would receive an unaffected *X* chromosome from a non-irradiated father. It could be predicted that matings between an irradiated female and a non-exposed male should produce relatively more daughters than sons. Conversely, a higher sex ratio (more boys than girls) would be expected from matings of irradiated males and unexposed females, since sons would receive the mother's unaffected *X* chromosome along with the irradiated *Y* (which carries relatively little genetic information), while daughters would receive one irradiated *X* chromosome as well as the unaffected *X* chromosome from the mother. Some studies, up to 1953, seemed to support these predictions, but when children born after that date were added to the data, the resultant sex ratios did not show significant deviations from the normal sex ratio. Any effects on sex ratio as a result of radiation exposure are no longer evident.

While the evidence of gametic mutations in survivors of the Hiroshima and Nagasaki populations is, at best, meager, it is clear that somatic mutations are increased by exposure to equivalent acute doses of radiation from atomic explosion experienced by the survivors of the bombing attacks on these two cities. This is further borne out by studies among the residents of Rongelap in the Marshall Islands. The native population of this island was accidentally exposed to fallout from thermonuclear device tests on neighboring islands in 1954. Over the years, a number of those who were children at the time of the test have developed tumors of the thyroid, and at least one young Rongelapese has died as a result of radiation exposure effects.

Genetic variation ultimately depends on mutation, at least for the origin of different alleles at any single locus, and mutation, from a long range perspective, provides a form of insurance against possible extinction. But over the short term, changes in the genetic material are more likely to have deleterious effects than to provide any immediate benefit to our species. Abundant evidence now exists to demonstrate that acute doses of high intensity radiation, whether from medical exposure or thermonuclear explosions, result in increased rates of somatic cell mutations. Exposure to radioactive materials or irradiation in medical procedures must be carefully weighed against the risks which would result if treatment were withheld. The medical profession has accepted the need to provide improved protection and withhold unnecessary diagnostic and treatment procedures in order to decrease exposure to mutagenic agents.

As yet, however, the rate of gametic mutations as a result of acute, intense exposure has been below that predicted on the basis of theoretical considerations, particularly among the victims of the bombings of Hiroshima and Nagasaki. It is probably unrealistic to have expected a drastic increase in the number of mutations to be observed in the first generation offspring of these unfortunate people. Gross changes, especially those involving changes from recessive to dominant alleles, may well have precluded successful fertilization or the development of the embryo to a stage sufficiently advanced to detect before an early spontaneous abortion occurred, even without the mother's awareness that she was pregnant. And, of course, mutations to recessive alleles, especially if these are already rare in the population, are unlikely to produce phenotypically recognizable traits until subsequent generations of breeding produce the appropriate homozygous recessive genotype combinations. We probably do not yet have accurate estimates of the amount of gametic mutation damage caused by nuclear explosions, but we already know that the effects on somatic cells have been deplorable.

Human populations are continuously exposed to chronic low-dose exposure as a result of low-level background radiation from radioactive materials in the environment, in materials we ingest, and from cosmic rays. Presumably, the lower mutation rates resulting from this source of chronic exposure would be sufficient to maintain genetic variability adequate for our survival as a species over time. It would seem the better part of wisdom, then, to continue our efforts to limit the development of unnecessary sources of mutagenic agents. In the next few decades we may all have to face a series of difficult decisions in choosing alternative forms of energy sources, some of them with mutagenic potential, to replace our dwindling supplies of gas, oil, and coal. Our decisions will affect not only us, but the well-being and health of our descendants for many generations to come.

PROBLEM SET: MUTATION

---

1. A certain genetic disease proved fatal to all infants born with the condition, all of whom died within twenty-four hours of birth. The disease was found to be caused by the presence of the single allele Z, and studies of stillborn babies indicated that embryos homozygous at this locus were so severely affected that none was ever liveborn. In 1960, two of every 1000 liveborn infants were found to have the condition, so it can be calculated that mutations at this locus must have been in excess of ________ per 100,000 gametes.

2. A certain population consisting of 1000 adults homozygous for the L allele formed a commune and moved into a valley beneath a leaky nuclear energy plant. After three generations, the allele frequency for this allele had receded to a value of 0.73 as a result of a one-way recurrent mutation rate which remained constant over the entire period. What was the value, $U =$ ________, of the mutation rate?
3. If a population with the allele frequencies of $p = 0.50$ and $q = 0.50$ was exposed to two-way recurrent mutation rates of $u = 0.05$ and $v = 0.10$, equilibrium should be reached at the point where $p =$ ________ and $q =$ ________.

REFERENCES AND RECOMMENDED READINGS

BLOOM, A.D., et al. 1970. Chromosome aberrations and malignant disease among A-bomb survivors. *American Journal of Public Health* 60: 641–644.

CONARD, R.A., et al. 1960 *et seq*. Medical survey of Rangelap people. Upton, New York: Brookhaven National Laboratory.

FINCH, S.C., and H.B. HAMILTON 1976. Atomic bomb radiation studies in Japan. *Science* 192:845.

GENTRY, J.T., E. PARKHURST, and G.V. BULIN, JR. 1959. An epidemiological study of congenital malformations in New York State. *American Journal of Public Health* 49: 1–22.

GRAHN, D. and J. KRATCHMAN 1963. Variation in neonatal death rate and birth weight in the United States and possible relations to environmental radiation, geology and altitude. *American Journal of Human Genetics* 15:329–352.

LEWIS, E.B. 1963. Leukemia, multiple myeloma and aplastic anemia in American radiologists. *Science* 142:1492–1494.

MEYER, H.H., et al. 1969. Investigation of the effects of prenatal X-ray exposure. *American Journal of Epidemiology* 89:619–635.

MILLER, R.W. 1969. Delayed radiation effects in atomic bomb survivors. *Science* 166:569–574.

STERN, C. 1973 *Principles of Human Genetics.* 3rd ed. San Francisco: W.H. Freeman Company.

CHAPTER 8

Genetic Drift and Breeding Isolates

For most of our history, at least until the last ten or fifteen thousand years, human populations have been limited in size and, to varying degrees, have been migratory in much of their daily life. A bountiful harvest of seasonal fruits or nuts, the successful ambush and kill of some large prey, or even a brief period of intensive collection of some localized plant or animal might have briefly interrupted the daily movements of small groups of foragers and hunters, but most of human life has been spent in nearly constant movement by small groups in an endless search for food. Only with the development of vessels and techniques for storing food and the practice of planting and harvesting food crops was it possible for humans to develop a sedentary life style. Even today, competent scholars argue as to whether intensive agricultural activities led to increased population size, or whether the growth of human populations was a necessary precondition for the kinds of extensive irrigation works which made agriculture possible in those parts of the world where some of the earliest evidence of large human settlements and high levels of human population densities are located.

There are many evolutionary consequences of this earlier pattern of life which lasted for millions of years of human history. Diseases which are spread through contaminated water supplies or from polluted sources, such as cholera or hepatitis, must have been virtually unknown until dense settlements arose with their attendant problems of human waste disposal and pollution of local water supplies. Most of the serious infectious diseases, such as bubonic plague or smallpox, which repeatedly decimated human groups until a few centuries ago in Europe, were not likely to have been maintained in small groups of hunters and gatherers living in relative isolation from one another. Even such contagious diseases as measles, mumps, or diphtheria, which still occur as childhood diseases among uninoculated children today, must have been infrequent, since the size and density of human host populations would have been too limited to maintain the causative organisms over time. Entire human groups lacking any acquired immunity against the more innocuous contagious diseases of European and Euroamerican children are still being found in the hinterland areas of the Philippine Islands and South America, and the response of these "virgin populations" to the introduction of such foreign microorganisms has too often been cataclysmic.

The genetic consequences of small size and isolation also occupy an area of special concern in the study of human evolution, since populations with these characteristics may be peculiarly subject to the operation of *random genetic drift,* sometimes referred to as "Sewall Wright effect." As described by the eminent mathematical geneticist, Sewall Wright, genetic drift referred to the random fluctuations in allele frequencies from one generation to the next, which would occur in a relatively isolated population of limited numerical size. Over time, the term has come to apply to several different genetically random processes affecting allele frequencies. One of these processes, that discussed by Wright, might be distinguished by the term *intergenerational gametic drift;* it refers to random deviations or nonsystematic fluctuations in allele frequencies as a result of sampling variations in the transmission of gametes from parents to an offspring generation. This process might be illustrated by comparison with a brief exercise in sampling procedures.

As anyone who has ever flipped a coin to see who is going to pay for coffee can tell you, a "50-50 chance" is no guarantee against penury. In ten consecutive tosses of an unbiased coin (that is, an unweighted coin with two different sides) there is a small, but real, probability that all ten tosses will end up as heads. The probability, or odds, of producing such a sequence can even be calculated from the appropriate statistical table, and it approximates a normal distribution curve. Even more appropriate to the issue of gametic sampling variance in a small, closed population are the probabilities of picking a representative small-size sample of marbles from a large collection. In a large jar of thousands of marbles, half of them red and the other half black (read "$p = 0.50$ and $q = 0.50$"), there is a probability of over 2 percent that a random sample of 50 marbles will include 40 percent or less (< 21) black marbles. It is also a property of sampling and probability that the larger the sample selected, the lower is the probability of getting such great variance in samples.

Humans are not marbles, of course, but chance also affects the transmission of gametes from one generation to the next in a human population. In a population of 50 adults with allele frequencies of $p = 0.50$ and $q = 0.50$, an offspring generation may deviate widely in the frequency of alleles as a result of sampling variance from the parental generation's gene pool. In about 2 percent of such samples, the resultant allele frequencies could be expected to be $p = 0.60$, or higher, and $q = 0.40$, or lower. But the offspring generation, when it begins reproducing, would be sampling from a gene pool with these different frequencies, $p \cong 0.60$ and $q \cong 0.40$. Sampling variance continues to operate in this group as their numbers remain limited and in the absence of the operation of other evolutionary forces, so we could expect to see the allele frequencies again subject to random fluctuation, or "drift." Through this process, a number of small populations, once identical in allele frequencies, might attain a remarkable degree of interpopulation variation.

Some students of evolution have argued that this process cannot have had evolutionary significance, since the effects of this form of genetic drift are limited to small and relatively isolated groups. Moreover, such geneticists as E.B. Ford have argued that selection is a more powerful and pervasive evolutionary force whose systematic effects can readily override the differentiation which random drift might produce.

It is quite true that the stronger such systematic pressures as selection, mutation, and gene flow are, the less effective will be the pressure of random sampling fluctuations in a small population. But this need not negate *any* effects by the random process. As we have already seen, most of human history has been lived by small, isolated populations, some of which ultimately grew in numbers. But the numerous modern descendants of some of those formerly small groups were derived from the limited gene pools of their few ancestors. Again, at least one important source of selective pressures affecting the gene pools of our ancient ancestors, certain forms of density-dependent diseases, were probably not operative until the most recent millenia of human existence, while gene flow between small, semi-isolated human groups in our prehistoric past must have been limited to contiguous populations. Finally, it should be pointed out that in our modern communities the numerous residents of a metropolitan city are often divided into a number of smaller groups by residence, ethnic identity, religious affiliation, and other cultural factors. These smaller enclaves, as discussed in an earlier chapter, often constitute the effective breeding populations, and the total numbers of such groups are, indeed, sufficiently limited to fulfill the conditions of small size and relative (breeding) isolation required for the operation of random intergenerational gamete sampling.

In consequence of these and other considerations, some anthropologists and geneticists have continued to accept the possibility that random gametic sampling has played a role in the creation and maintenance of human variation. But, because the process is an indeterminate one, it has been notoriously difficult to amass clear-cut evidence of its operation. One approach to this problem has been the attempt to detect allele frequency differences between parental and offspring generations in a single breeding population. Hulse (1960), for example, found significant differences in *ABO(H)* blood group frequencies between the younger ($<$ 16 years of age) and the older generations of Indians living on the Hupa Indian reservation of northwestern California. Although the residents on this reservation include a number of individuals with mixed parentage, representing other non-Hupa Indians as well as Europeans and Euroamericans, statistically significant differences in *ABO(H)* allele frequencies were not found to be associated with degree of Indian ancestry or with tribal affiliation. Hulse concluded that random genetic drift and mixture, acting together, were affecting the observed shifts in allele frequencies among the inhabitants of the reservation.

Beginning in 1952, Bentley Glass and his colleagues investigated the frequencies of three blood group systems and several visible genetic traits in the Older Order German Baptist Brethren ("Dunkers") of Franklin County, Pennsylvania. This breeding isolate was formed from some of the descendants of a religious sect that originally came to Pennsylvania from West Germany in 1719. The group was joined by other migrants in later years, but in 1881 the sect divided into three groups, with the more religiously orthodox members remaining to constitute the present Old Order Brethren. Glass compared the allele frequencies for the *ABO(H), Rh* and *MN* blood group systems between three generations of Dunkers and found statistically significant differences by age group at the *MN* locus. Although he argued that this striking difference, shown in Table 8-1, demonstrated the operation of genetic drift, others have challenged this claim by pointing to the rather high rate of emigration characteristic of this group in which about one-fourth of all children leave upon reaching adulthood.

More commonly, random genetic drift in the form of intergenerational gamete sampling variance, is evoked to explain allele frequency differences between genetically related, small and relatively isolated populations where the evidence is insufficient to demonstrate that selection or gene flow can account for the observed variations. For example, differ-

TABLE 8-1 Intergenerational Differences in MN Allele Frequencies Among the Dunkers

Age Group	*Allele Frequencies* *M*	*N*
3–28 years old	0.735	0.265
28–55 years old	0.66	0.34
56 years or older	0.55	0.45

From B. Glass. 1956. *On the evidence of random genetic drift in human populations.* American Journal of Physical Anthropology, Volume 14. Reprinted by permission of the Wistar Institute Press.

ences in allele frequencies at several of the blood group system loci among various groups of Eskimos (Laughlin, 1950) and in island populations of Oceania (Giles, 1973) have been ascribed to the effects of intergenerational gametic sampling variations; the populations of both the Arctic regions and of Pacific islands present ideal conditions for the operation of genetic drift in relatively small groups occupying homogeneous environmental settings.

Giles and his colleagues (1970) studied blood group allele frequency differences among the New Guinea native populations of the Markham Valley and of the Eastern Highlands. The populations belonging to each of the two groups were genetically related, occupied a similar environment, and shared common cultural patterns and techniques of exploiting the environment, but little gene flow took place among the populations of either group. Giles concluded that the observed extensive variations in allele frequencies among populations belonging to each of the two groups could best be attributed to the operation of random genetic drift.

On the island of Yap in western Micronesia, the natives (numbering about 4000 in 1966) of this small island, which measures less than 38 square miles in area, were divided into ten residential groups, or Districts. Earlier studies had indicated that intermarriage between members of different districts was uncommon, so each district represented a relatively isolated breeding population. On so small an island, ecological differences among districts were slight, and cultural, including technological, differences were minimal. When allele frequencies at the *ABO(H)* blood group locus are examined (Table 8-2), variations of the magnitude distinguishing racial groups in humans are found. These results are, again, highly suggestive, but not conclusive proof of the role of random factors in the microevolution of the Yap Islanders.

Interpretations which rely on intergenerational gametic sampling fluctuations as an explanation by excluding the probable operation of

TABLE 8-2 ABO(H) Blood Group Allele Frequencies, by District, Yap Islands, 1966.

District	*ABO(H) Blood Group System Allele Frequencies*		
	p_A	q_B	r_O
Dalipebinau	0.195	0.155	0.657
Fanif	0.137	0.137	0.707
Gagil	0.154	0.102	0.751
Giliman	0.258	0.072	0.688
Kanifay	0.000	0.194	0.764
Map	0.138	0.072	0.800
Rull	0.189	0.064	0.752
Rumung	0.058	0.112	0.837
Tomil	0.030	0.077	0.887
Weloy	0.300	0.096	0.620
Totals	0.153	0.100	0.748

After Hainline, 1966.

other evolutionary forces often fail to account for the possibility that another random process, founder effect, may be responsible for the origin of those differences that intergenerational random variation has, at best, merely accentuated. *Founder effect* refers to the phenomenon in which small colonies of founders may not accurately represent the entire range of variation of the ancestral gene pool. For example, from an ancestral group with allele frequencies of $A = 0.60$ and $a = 0.40$ at a particular locus, a group of founders may leave to form a colony elsewhere. This founding group, by chance alone, may be composed of a few individuals all of whom have the genotype *AA*, and become the ancestors of subsequent generations of a new population entirely lacking the *a* allele. Despite the actual genetic relationship, the two populations in subsequent generations would appear quite dissimilar in allele frequencies.

In one case, that of the native populations of the Wellesley Islanders (in the Gulf of Carpenteria off northern Australia), historical, linguistic, and cultural evidence argues strongly for the operation of both kinds of random genetic processes in the genetic diversification of related populations. The Wellesley Islanders include the aboriginal populations of the islands of Bentinck, Mornington, Forsyth and Denham Islands, numbering all together 448 people in 1960. In respect to linguistic differences, as well as to differences in the material culture and social organizational patterns of these groups, the Bentinck Islanders appear to be the most isolated from native mainland Australian contact and from contact with Europeans or Malayans. Sharp differences are found in allele frequencies (Table 8-3)

TABLE 8-3 ABO(H) and Rh Blood Group System Allele Frequencies Among Natives of the Wellesley Islands

		Allele Frequencies						
		ABO Blood Group System			*Rh Blood Group System*			
Population	*N*	p_A	q_B	r_O	R^1	R^2	R^o	R^z
Bentinck	42	0.000	0.244	0.756	0.524	0.059	0.417	0.000
Mornington & Forsyth	67	0.078	0.000	0.922	0.656	0.260	0.068	0.016

After Simmons, Tindale, and Birdsell, 1962.

at the *ABO(H)* and *Rh* blood group systems loci between the Bentinck Islanders and the natives of Mornington and Forsyth Islands. Unlike the natives of Mornington and Forsyth Islands, or of native populations in neighboring mainland Australian areas, the Bentinck Islanders are totally lacking in the *A* allele and have the highest frequency of the *B* allele (0.224) found among tested northeastern Australian aborigines.

As the authors of the study point out, the two major island groups, Bentinck and Mornington-Forsyth, show allele frequency differences as great as those found between major races of the world despite cultural and geographical evidences of former relationship. Historical records, archaeological remains and genealogical records indicate that no alien genetic penetration is likely to have occurred into the Bentinck population, while contact with mainland or other natives was slight and limited to the residents of Mornington Island. However, in the three years before the Bentinck Islanders' 1948 removal to Mornington Island, 40 percent of this population died as a result of loss of life at sea (when trying to raft across to nearby islands) and from killings among a group wracked by a disastrous prolonged drought. Surely, natural disasters in the past must have operated to periodically reduce the total population of Bentinck Island, producing a "bottleneck" effect through which the population must have repeatedly gone. This bottleneck effect, the repeated decimation of the native population of Bentinck Island, produces a kind of founder effect, leaving a few randomly selected founders to become the ancestors of subsequent generations. There is probably no need in this case to attempt to distinguish founder effect from intergenerational gametic sampling effects, for both have undoubtedly played a role in the continuing creation and maintenance of genetic diversification among these northeast Australian island populations.

In another case, Roberts (1968) has provided a convincing argument that bottleneck effect has played a major role in the creation and maintenance of existing levels of genetic polymorphism found in the population of Tristan da Cunha, a remote island in the south Atlantic. This island, first discovered by Europeans in 1506, was the site of several explorations and abortive colonization attempts over the following few hundred years. But in 1816 a military garrison was established by English authorities and then withdrawn, leaving behind a corporal of Portuguese ancestry, born in South Africa, and his wife, who were joined thereafter by a few other brave pioneers. In 1817, this small group signed an agreement regarding the goals of the colony and certain basic principles around which this "social experiment" would be organized. By 1826, five adult males resided on the island, but only one of these was married, so a number of prospective wives was brought into the colony, and from time to time over the following years, other new colonists joined the group.

By 1855 the island population numbered some 103; 25 of these emigrated to the United States in 1856, and another 45 islanders left for South Africa in 1857, reducing the island population to 33. The island population grew again from these small numbers to 106 in 1885, but 15 adult males were drowned at sea, and in the next few years a number of their widows and orphaned children left the island, until only 59 people were left on Tristan da Cunha in 1891. The effects of these two drastic population reductions were to eliminate from the modern population all genes derived from a portion of the ancestors and to increase the relative proportion of the contribution to the modern gene pool made by some of the remaining ancestors.

But if repeated bottlenecks over time have accidentally created a series of small founder groups in some populations, other human colonies have been intentionally founded by small groups of colonizers, not only on scattered islands or in isolated hinterland areas, but also as self-delimited enclaves among larger populations in modern countries. Among the latter category are found the members of certain religious sects, such as the Dunkers described earlier in this chapter, but also such religious communities as the Hutterites and the Amish of North America. Until a few years ago, about 45,000 Old Order Amish lived in the United States, with over 80 percent of these located in the states of Indiana, Ohio, and Pennsylvania. The sect was founded in 1693 in Switzerland as a splinter group from the Mennonite Church, and most of the Amish living today in the northeastern United States are descendants of immigrants who came to this country between 1720 and 1770. Until recent years, the Amish were a self-defined religious sect, noticeable for their style of "plain dress," the use of horse and buggy for transportation, and their continuing use of a South German dialect. Although few Amish ever left the group, virtually no one immigrates into the colonies, and sect principles constrain many forms of participation in the larger society of modern America.

In each of six local areas of the northeastern United States where Amish colonies are found, roughly 80 percent of all families share one of ten family names. For example, as shown by McKusick and his colleagues (1964), 81 percent of the 1106 Amish families residing in Lancaster County, Pennsylvania, had one of eight surnames; 85 percent of the 238 Amish families in Mifflin County, Pennsylvania, had one of a set of eight different surnames. As genealogies available in each community indicate, the Amish colonies of today are composed primarily of the descendants of a very few ancestral founders.

The genetic consequences of this pattern of colonization are singularly impressive. In each of four different Amish communities, at least one recessive genetic disorder is found in relatively high frequencies. Less than fifty cases of one such condition, Ellis-van Creveld syndrome (a form of dwarfism often associated with a congenital heart abnormality) had been reported in the medical literature prior to McKusick's study of the Amish. But forty-three cases of the condition were found among Amish living in eastern Pennsylvania, and all fifty-two parents of afflicted Amish individuals can be traced back to one common ancestor. Similarly, several other genetic disorders which are only rarely found in most populations appear at respectable frequencies among the Amish and can be traced back to one or a few ancestors in each of the Amish communities where the trait is present. Clearly, in these semi-isolated colonies, each descended from a few ancestors, the resultant level of inbreeding has uncovered rare recessive alleles which appear in the phenotypes of the homozygous progeny of genetically related spouses.

A third form of random genetic pressures affecting the genetic microdifferentiation of small human populations has been described by investigators studying the Yanomama of southern Venezuela and northern Brazil. This tribal group, numbering some 10,000, live in over 100 small villages. New villages are most commonly formed as a result of the fissioning of an existing village; the founders of the new group tend to be genetic relatives, such as two brothers, who, accompanied by their wives and children, leave the older village to form a new settlement. Thus, each new colony is likely to include members with a higher coefficient of inbreeding than is characteristic of the original village. This *lineal effect,* a genetically random but socially systematic process, produces a high degree of genetic differentiation among Yanomama villages. Chagnon (1972) shows that the fissioning process, socially non-random founder selection, can lead almost instantaneously to high levels of genetic divergence. Differences among villages may then be substantially reduced over time, as dominant patterns of migration and gene flow counter the effects of fissioning, with the surprising result that genetic differences are less between Yanomama villages separated over longer periods of time than between recently separated parental and offshoot villages.

A fourth possible form of genetic drift, *headman effect,* has also been reported by investigators working among relatively unacculturated tribal groups of South America. Among the Xavante of the Brazilian Mato Grosso, for example, Salzano and his co-workers (1967) found that chiefs of villages and heads of clans practiced polygyny to a greater extent than did other adult males in the three villages that were studied. While slightly over 40 percent of all married men were engaged in polygynous unions, headmen had more wives (four or five each) and produced on the average more children (9.5 average). Furthermore, a comparison of the mean number of children and the variance in number of children born to Xavante men reveals that the reproductive success of these few highly polygynous males is offset by the very low performance of a larger number of socially undistinguished males. If we assume that headmanship in this group of hunting-gathering people who practice an incipient form of horticulture is not determined by genetic factors, as seems to be the case, then the social determination of male status directly affects the genetic diversity of Xavante groups. Again, systematic social factors constitute a form of genetically random pressure on the gene pool of these populations. Nor is headman effect confined to the Xavante; for, among the Yanomamo, among whom headmen are also determined by their political and military abilities, village headmen also make relatively much larger contributions to the gene pool than do their fellows. As noted in discussing assortative mating patterns, Chagnon et al., (1970) report that in one Yanomama village, two headman sired twenty-eight children between them, or roughly one-fourth of the entire population of the village.

Essentially, qualitative approaches to the study of genetic drift in human populations provide at best suggestive interpretations of observed phenomena. A more rigorous approach entails the use of mathematical models which enable us to quantify the necessary conditions which must prevail if random processes are to operate as an evolutionary force and to predict the limits, but not the directions, beyond which random fluctuations are unlikely to occur. These models help to answer basic questions as to the intensity of natural selection required to override the effects of random factors on genetic differentiation, the amount of gene flow necessary to maintain genetic homogeneity among groups of related populations, and even help us to calculate the maximum range of expectable random fluctuations in populations of different sizes and with different population structures.

In any human group, the total number of the population includes some proportion of adults in the reproductive age groups, as well as others (children, post-menopausal women, aged men) who are not reproductively effective. In a population of 1000, one might find at a given time that perhaps 45 percent of the whole group are aged 0–14 years; 30 percent are aged 15–44; and the remainder, 25 percent, are aged 45 years or older. On the whole, females past age 44 contribute only very slightly to the total number of births in any population, and in many human societies, males beyond this age father relatively few children. Accordingly, only 30 percent, or 300, of this group constitute the potentially reproductively active segment of the total population. We might refer to this as the number of potential breeders (N_{pb}). Ideally, the gene pool of this generation would contain all the genetic variation present in the genomes of these 300 people and this variation would be maintained from generation to generation if the size of the population remained constant, everyone mated at random, and each couple produced an equal number of children who survived to reproduce.

In fact, all these ideal conditions are not met by human populations, so the *effective population size* (N_e) is inevitably lower than the number of potential breeders (N_{pb}). Suppose that the 300 adults in this group consist of 100 males and 200 females. If lifelong monogamous unions are the only permitted form of mating, the genetic variation represented in the genomes of the surplus females will not be transmitted at all to the offspring generation. Even if this population practices polygyny (the marriage of a man simultaneously to two or more wives), the genetic contribution of reproductively successful males will be proportionately over-represented in the offspring generation. The effects of inequality in sex ratio on the effective population size can be calculated from a simple formula,

$$N_e = \frac{4N_f N_m}{N_f + N_m}$$

where N_f refers to the number of females and N_m refers to the number of males.

In a population with a sex ratio of 100 (that is, with equal numbers of reproductively active males and females), the effective population size (N_e) is identical with the total number of potential breeders. In the hypothetical example being considered, we need only substitute the values of 150 for males and 150 for females:

$$N_e = \frac{4(150)(150)}{150 + 150} = \frac{4(22{,}500)}{300} = \frac{90{,}000}{300} = 300.$$

But, if the number of males is 100 and the number of females is 200, the value of N_e is reduced, as shown:

$$N_e = \frac{4(200)(100)}{200 + 100} = \frac{4(20{,}000)}{300} = \frac{80{,}000}{300} = 266.7.$$

Similarly, the value of N_e is reduced as a result of inbreeding or cyclic variations in population size. But most critical to our consideration of genetic drift is the effect of variance in the number of offspring produced.

In any human population, couples produce different numbers of children, some women bearing no children at all, while other couples may produce up to ten or more children. In a population which is stationary, the numbers and proportions of various age groups neither increasing nor decreasing, the average number of children produced need only be sufficient to replace both parents and to compensate for prevailing death rates. In some modern societies, with low mortality rates among infants, children, and younger adults, the average number of offspring required to maintain the population at a constant size and age structure is slightly in excess of 2. In groups with higher mortality rates, especially among children and sub-adults, this number is larger. From the viewpoint of our interest in genetic diversity, it is not the average number, but the variation in number of offspring produced by all couples that is important. This can best be illustrated by calculating the mean (average) and variances (s^2) in the number of offspring for two sets of data (Table 8-4). Patently, in the one population there is considerable variance in the number of children produced; thus, variance in the extent to which some couples contribute to the gene pool of the next generation in comparison to other couples, although the mean, or average, number of children produced by couples in both cases is identical (3).

The effect of variance in the number of offspring on the effective population size can be estimated by the use of the following equation:

$$N_e = \frac{4N-2}{2 + s_k 2}$$

where N is the number of adults of reproductive ages (potential breeders), and $s_k{}^2$ refers to the variance in the numbers of offspring born to these adults. Johnston (1973) has calculated that the effective population size of the Cashinahua Indians of Peru, numbering a total of 206 in 1966, is only 68—although 87 adults of reproductive age are found in this population. From this information he is able to further calculate that the expected maximum deviation at one blood group locus (Kidd) due to

TABLE 8-4 Calculation of the Mean and Variance in Number of Offspring, Two Population Samples

Couple	*Number of Children* X	*Mean* $\bar{X}$	$(X - \bar{X})$	$(X - \bar{X})^2$
		Case #1		
A	2	-3	-1	1
B	4	-3	+1	1
C	6	-3	+3	9
D	4	-3	+1	1
E	2	-3	-1	1
F	0	-3	-3	9
			0	22
		Total N of children = 18 Number of couples = 6 Mean N of children = 3 $s^2 = 22/5 = 4.4$		
		Case #2		
A	3	-3	0	0
B	3	-3	0	0
C	3	-3	0	0
D	3	-3	0	0
E	3	-3	0	0
F	3	-3	0	0
			0	0
		Total N of children = 18 Number of couples = 6 Mean N of children = 3 $s^2 = 0/5 = 0.0$		

sampling fluctuation in one generation is 0.04 for the Jk^a allele. This is not to say that such a change actually occurred, nor that the direction of any such change could be predicted, but that a change of this magnitude *could* occur as a result of random factors operating on a population of this effective size.

It follows, then, that the total size of the population provides only a general indicator of the existence of those conditions necessary for genetic drift to operate. Genetic variation may be reduced by cyclic changes in population size, by inbreeding, by unequal sex ratios, and by changes in variance of the number of offspring. All these factors reduce the effective population size (N_e), thus enhancing the opportunity for random factors to operate on the allele frequencies of human populations. Various estimates of the effective population size for a number of populations approximate one-fifth to one-third of the total population; so, many of those human groups studied by anthropologists fulfill the necessary condition, small size, required for the operation of genetic drift. Since the effect of drift per generation may be estimated by the formula, $pq/2N$, it is possible to briefly compare the possible consequences of this random evolutionary force on populations with different values for N_e (Table 8-5). As indicated, the predicted effects of drift are greatest in populations with small effective breeding size and maximum genetic diversity ($p = q = 0.05$).

Populations with limited effective population size are not immune to the operation of other evolutionary forces and, as already noted, some authorities have argued that the overriding effect of systematic factors (selection, gene flow, mutation) is likely to make drift a negligible factor in the maintenance of genetic variation among human populations. According to Wright, the necessary conditions for the operation of drift (small size of population) are met when N is equal to or less than one half the value of the coefficient of selection, the mutation rate or the rate of gene flow ($N = \leqslant 1/2s, \leqslant 1/2u, \leqslant 1/2m$). Unfortunately, it is usually impossible to determine whether these values have been exceeded in the past histories of specific populations, and we are usually forced to rely on the absence of any detectable evidence for selective pressures in evoking drift

TABLE 8-5 Estimated Effects of Genetic Drift ($\frac{pq}{2N}$) per Generation on Model Populations at Differing Values of N_b, N_e, p and q

N_b	N_e		*Maximum Change*
		Case #1, p = q = 0.50	
30	6		0.0208
	15		0.0083
	24		0.0052
300	60		0.0021
	150		0.0008
	240		0.0005
		Case #2, p = 0.80, q = 0.20	
30	6		0.0133
	15		0.0053
	24		0.0033
300	60		0.0013
	150		0.0005
	240		0.0003

as an explanation for observed genetic variations among modern human populations.

In recent years, computer simulations have been used to devise a probabilistic range of values for allele frequencies in model populations. These model values can then be compared to real data to test whether observed differences can be explained as resulting from the operation of random factors. In one such study, Cavalli-Sforza (1969) and his associates proposed to test the hypothesis that drift should be more effective in small, isolated populations than in large, less remote groups. When data on genetic variation among village populations in the Parma Valley of Italy were analyzed in this manner, it was found that intervillage variation increased as village population density and geographic isolation increased. Moreover, while natural selection, if operative, would have resulted in changes in some gene frequencies, but not on others, observed variations between and among villages were about the same for any gene studied—thus eliminating natural selection as a satisfactory explanation for the observed variations.

Finally, artificial populations of twenty-two "villages" were generated in a computer and assigned values for the *ABO(H), MN* and *Rh* blood group system allele frequencies. Using randomizing procedures, these populations were exposed to simulated births, deaths, and marriages for a number of generations. The variations among the simulated villages were found to match observed variations among real villages, thus supporting the interpretation that genetic drift suffices to explain the pattern of genetic diversity found in Parma Valley villages.

It appears reasonable to consider genetic drift, in any of its various forms, as an evolutionary force of potential importance in the creation and maintenance of human genetic diversity. But, because of the nonsystematic nature of its operation, it is virtually impossible to demonstrate that drift is the unequivocal and single force responsible for genetic differentiation. Attempts to prove that drift is at least a primary cause of differentiation have usually attempted to show that the observed results cannot reasonably be attributed to the operation of other evolutionary forces and that the magnitude of observed differences falls within the range which mathematical models would predict could be expected if genetic drift were operative on a relatively isolated population of small effective population size. While these mathematical models, enhanced by the versatility and sophistication that computer simulations can provide, become ever more comprehensive, "proof" still consists essentially in the congruence between predicted results and observed data.

This problem is not unique to the study of genetic drift; for, with one notable exception (sickle hemoglobin), examples of natural selection in human populations have also had to rely on consistency between expected and observed results. But in the case of systematic evolutionary pressures, natural selection and gene flow, both the magnitude and the direction of change can be predicted, thus improving the reliability of predictions and increasing the points of congruence. Nonetheless, because so much of human evolutionary history has been experienced under the very conditions (small population size and semi-isolation) that permit the operation of genetic drift, our understanding of human evolution and genetic diversity require continuing study of genetic drift. As human groups erect cultural barriers, replacing those geographical constraints which formerly separated one human society from another, small breeding isolates continue to provide us with a reservoir of genetic diversity within this modern polytypic species to which we belong.

PROBLEM SET: GENETIC DRIFT

---

1. In the mythical land of Xingpu, the ruling monarch was offended by the presence of all men who suffered from partial color-blindness, an X-linked trait. He decreed that all young adult males who suffered this affliction (constituting ten of the total 100 men belonging to this age group) must forthwith leave his country, taking with them an equal number of young women, all of whose fathers had suffered from the same affliction. These hapless exiles left Xingpu, tears and sad farewells accompanying their departure, and sailed eastward until they reached an uninhabited island, where they settled and began a new way of life. Some twenty-five years later, a young anthropologist visited Xingpu and, hearing of this tale, was asked by the aging monarch if there were likely to be any men now living on the island with normal color vision. The scientist predicted that under ideal circumstances the frequency of unaffected males should be about ________.
2. In a number of societies of hunters and gatherers, intentional infanticide was practiced against females and may, in the past, have amounted to the killing of as much as 50 percent of all female babies ever born. Assuming that the secondary sex ratio was 100 and no sex differential in mortality existed between males and females not removed through infanticide, what effect would this practice have on the value of N_e? ________

REFERENCES AND RECOMMENDED READINGS

CAVALLI-SFORZA, L.L. 1969. Genetic drift in an Italian population. *Scientific American,* 221:30–37.

CHAGNON, N.A. 1972. Tribal social organization and genetic microdifferentiation. In *The Structure of Human Populations.* G.A. Harrison and A.J. Boyce, eds. Oxford: Clarendon Press, pp. 252–282.

CHAGNON, N.A.; J.V. NEEL; L. WEITKAMP; H. GERSCHOWITZ; and M. AYRES 1970. The influence of cultural factors on the demography and pattern of gene flow from the Makiritare to the Yanomama Indians. *American Journal of Physical Anthropology,* 32:339–350.

GILES, E. 1973. Population analyses in Oceania. In *Methods and Theories of Anthropological Genetics.* M.H. Crawford and P.L. Workman, eds. Albuquerque: University of New Mexico Press, pp. 389–402.

GILES, E., S. WYBER and R.J. WALSH 1970. Microevolution in New Guinea: additional evidence for genetic drift. *Archaeology and Physical Anthropology in Oceania,* 5: 60–72.

GLASS, B. 1956. On the evidence of random genetic drift in human populations. *American Journal of Physical Anthropology,* 14:541–555.

GLASS, B.; M.S. SACKS; E.F. JAHN; and C. HESS 1952. Genetic drift in a religious isolate: an analysis of the causes of variation in blood group and other gene frequencies in a small population. *The American Naturalist* 86:145–159.

HAINLINE, J. 1966. Population and genetic (serological) variability in Micronesia. *Annals of the New York Academy of Sciences,* 134: 639–654.

HULSE, F.S. 1960. Ripples on a gene-pool: the shifting frequencies of blood-type alleles among the Indians of the Hupa Reservation. *American Journal of Physical Anthropology,* 18:141–542.

JOHNSTON, F.E. 1973. *Microevolution of Human Populations.* Englewood Cliffs, N.J.: Prentice-Hall, Inc.

LAUGHLIN, W.S. 1950. Blood groups, morphology and population size of the Eskimos. *Cold Spring Harbor Symposium Quantitative Biology,* 15:165–173.

McKUSICK, V.A.; J.A. HOSTETLER; J.A. EGELAND; and R. ELDRIDGE 1964. The distribution of certain genes in the Old Order Amish. *Cold Spring Harbor Symposium on Quantitative Biology,* 29:99–114.

ROBERTS, D.F. 1968. Genetic effects of population size reduction. *Nature,* 220:1084–1088.

SALZANO, F.M.; J.F. NEEL, and D. MAYBURY-LEWIS 1967. I. Demographic data on two additional villages: genetic structure of the tribe. *American Journal of Human Genetics,* 19:463–489.

SIMMONS, R.T.; N.B. TINDALE, and J.B. BIRDSELL 1962. A blood group genetical survey in Australian aborigines of Bentinck, Mornington and Forsyth Islands, Gulf of Carpenteria. *American Journal of Physical Anthropology,* 20:303–320.

CHAPTER 9

Natural Selection in Human Populations

If human populations consisted of genetically identical individuals, evolution—changes in the allele frequencies of a population's gene pool—could not take place. Thus, the necessary precondition for human evolution is the existence of genetic polymorphisms. It scarcely seems necessary to provide a lengthy documentation of the existence of polymorphisms in human populations, for McKusick (1971) has provided us with a catalogue of thousands of human genetic traits, and the list grows longer every day. Harris's studies (1959, 1966) established that 97 percent of the members of an English population were heterozygous for at least one of ten loci directing the production of certain enzymes and proteins, while polymorphisms were found for about one-fourth of the enzymes of eighteen enzyme systems studied in European and African black populations. Other estimates suggest that each of us is probably heterozygous at 3200 loci or more; yet, presumably, many more genetic polymorphisms exist in human populations than have even yet been detected with the tests currently available. The presence of genetic polymorphisms in human populations now thoroughly, if incompletely, demonstrated, certainly provides adequate materials on which evolutionary forces can operate.

It must stand to the everlasting, independent credit of Charles Darwin and Alfred Wallace that each recognized variation as the essence, not the imperfection, of biological reality, and each proposed that the differential success of some variants under prevailing environmental conditions constituted evolution through natural selection. Neither man comprehended the genetic basis of that biological variation; the first edition of Darwin's opus appeared several years before Gregor Mendel's pioneering essay on genetics was published. Not surprisingly, Darwin described natural selection in terms of the "survival of the fittest," while we now recognize that the measure of fitness is not mere survivorship but the degree to which individuals with certain genetic traits produce relatively more reproductively successful offspring than do individuals lacking these traits. Reproduction, not merely survivorship, is the crux of natural selection, and differential mortality is of importance as it eliminates or reduces the contributions made by possessors of some genotypes to the gene pool of succeeding generations.

As we have already seen, changes in allele frequencies can occur when mixture takes place between members of two formerly distinctive populations, when random sampling fluctuations occur, or, at least briefly, when new genetic materials are created through mutation. Changes can also occur even in numerically large, genetically isolated populations, through natural selection; but, uniquely, selection provides the primary mechanism through which environmental influences affect the genetic constitution of a population. This is accomplished through the differential reproduction of individuals whose genotypes confer a phenotype which enables their possessors under prevailing environmental conditions to survive and produce relatively more offspring, who themselves attain adulthood. Since environments change over time, this capacity to adjust to varying environmental pressures through changes in the gene pool provides a means of *genetic adaptation,* changes in the genetic constitution which enhance the population's biological success. The process can perhaps best be illustrated through an example of natural selection involving the maintenance of hemoglobin polymorphisms in certain African populations.

Several decades ago, it was suggested that individuals might differ in their susceptibility to malarial infection. One form of malaria, produced as a result of infestation by *Plasmodium falciparum,* is widespread in tropical areas; historical accounts indicate that it has been present since prehistoric times in parts of Africa and the Mediterranean basin. In sub-Saharan African populations, the frequency of certain hemoglobin variants, particularly sickle hemoglobin, parallels the distribution of malaria. But individuals with the homozygous genotype *SiSi* suffer from a fatal disease, sickle cell anemia (sicklemia); so the *Si* allele is constantly subject to removal from the gene pool of each of these populations. The continuing presence of this allele at the relatively high frequencies recorded among these groups cannot be explained by mutation alone, since unrealistically high rates of mutation would be required to account for the observed frequencies of the allele.

In 1954, A.C. Allison reported that individuals who were heterozygous at this locus were more resistant to malaria than were homozygotes (*AA*) for normal adult hemoglobin. He conjectured that selection operated against the *AA* homozygotes through the increased mortality from *P. falciparum* infestations which they experienced in contrast to the *ASi* heterozygotes. Selection also operated against homozygotes for the sickle allele, since most individuals with *SiSi* genotype died of sickle cell anemia before reaching adulthood. Subsequently, this hypothesis has been supported by several facts: children with the heterozygous genotype have lower levels of infestation by *P. falciparum* than children with the *AA* genotype; studies on volunteers of both genotypes who were experimentally infected with the parasite revealed lower rates of infestation in heterozygotes; and statistical studies have shown that heterozygotes in a number of these populations experience a lower frequency of fatal malarial infections.

We might, then, reconstruct the genetic history of the sickle hemoglobin allele among West African populations along the lines suggested by Frank Livingstone (1958). Archeological evidence suggests that African black populations of hunters and gatherers occupied most of West African tropical rain forest areas until the introduction of iron tools made it possible to productively cultivate these soils, particularly to grow the indigenous yam plant. Linguistic data support the notion that Bantu peoples, moving out from a central homeland in the Benue River valley of Nigeria, spread through central and southern Africa. This movement of ironworking, yam-growing peoples must have occurred after the introduction of iron working into western Africa, perhaps at the time of the beginning of the Christian era. Other peoples, raising different crops, appear to have moved at later dates into western Africa, and some of these peoples probably mixed with the earlier arriving yam-growers, while some of the yam-growers subsequently added other crops to their repertoire. Even today, however, the frequency of the sickle hemoglobin allele parallels the distribution of yam-growing peoples in West Africa, being lowest among hunting groups where both the allele and agriculture are only recently beginning to spread.

The relevance of agricultural activities to the history of the sickle hemoglobin allele becomes apparent in the light of further information about the nature of malaria and the mode of its spread. The parasite, *Plasmodium falciparum*, is transmitted to humans by mosquitos, particularly by the mosquito *Anopheles gambiae.* But *A. gambiae* normally cannot breed in running water or in shaded, brackish, alkaline, or polluted water. Only when humans cleared forest areas in the course of agricultural activities did the breeding sites for *A. gambiae* become numerous, as unshaded or stagnant pools of water, human refuse areas, and open swamps began to appear. Moreover, populations of agriculturalists occupying settled village sites provided both preferred habitations for mosquitos and a population size and density suitable to the continued reinfection of human hosts by *P. falciparum.* Even the reduction of the number of large game animals in these agricultural areas probably contributed to the spread of malaria among humans, as alternative mammal hosts for the organism become less common.

Presumably, the mutation for the sickle hemoglobin allele has occurred repeatedly in the course of human history, and the allele may well have been present at very low frequencies in populations of ancient West African hunters and gatherers, carried by occasional individuals with the heterozygous genotype. As iron-working and yam-growing Bantu populations moved into tropical forest areas and cleared the land for the growing of their crops, local populations of the *A. gambiae* mosquito became larger and fed on a growing number of settled villagers. The constant reinfestation of a dense human population led to the persistence of holendemic malaria, and under these circumstances, the heterozygous carrier of the sickle hemoglobin allele was at a selective advantage. The sickle hemoglobin homozygote (*SS*) continued to be under severe selective stress, while homozygote *AA* children died more often from malaria than did heterozygous children. But heterozygotes contribute both the *A* and *Si* alleles to the gene pool, thus maintaining genetic diversity in the population. The end result of natural selection in this case has been the maintenance of a *balanced genetic polymorphism*—the continuing presence of two alleles at a locus at frequencies above those which can be explained by mutation alone, in this case as a result of selection against both homozygotes.

Soon after the advent of DDT and new anti-malarial drugs and treatment, malaria seemed to be coming under human control; optimistic health officials once predicted the eradication of malaria. If this were to happen, the selective advantage once held by heterozygotes in these areas would disappear. But the picture is now far from being so cheerful, and new forms of DDT-resistant mosquitos have already appeared. At least in the case of American Blacks, however, the sickle hemoglobin trait appears to be lower in frequency than expected (unless selection is now operating against both the *ASi* and the *SiSi* genotypes in their new environment). Without the heterozygote advantage experienced in areas of endemic falciparum malaria, the sickle hemoglobin allele has been decreasing in frequency among Blacks in non-malarial areas of the New World. Indeed, it has been argued that the heterozygote, who produces some portion of hemoglobin which sickles under oxygen stress, may now be subject to an additional form of selection in smoggy urban areas of the industrialized parts of the New World where industrial pollutants contaminate the air.

This example, probably the best studied case of natural selection in human populations, exemplifies a form of selection known as *diversifying selection* in which genetic variation is maintained–in this case by means of the selective advantage of the heterozygote–in an environment in which endemic falciparum malaria prevails. But the case illustrates a more general property of natural selection–the dynamic quality of its operation. New mutations and associations of alleles into new genotype combinations appear and reappear in human populations, and their fate at a given time and under prevailing conditions is specified by the reproductive advantage or disadvantage conferred by the genotype on the individual living in that environment. Presumably, the sickle hemoglobin allele had appeared at various times in West African populations; but when the allele frequencies reached a level at which individuals with the *SiSi* constitution were born, the allele was subject to removal from the population as children with sickle cell anemia died before reaching reproductive maturity. Only when agriculturalists created the conditions in which endemic malaria was present did the sickle hemoglobin allele confer a reproductive advantage to the individual heterozygous at this locus. Where the selective agent, falciparum malaria, was not present, selection operated against the allele, at least when present in the homozygous combination. It is precisely this dynamic relationship between environmental agents of selection and the composition of the gene pool of a population which provides the primary mechanism of genetic adjustment to changing external conditions.

This is a far cry from older notions which envisaged some perfect genetic composition toward which populations inexorably moved through a process of automatic, even mechanical, rejection by natural selection against any new mutations or genotype combinations which might occasionally appear. Rather, in modern perspective, mutations and novel genotypes continue to appear, and recent studies suggest that some of these mutations may even be "neutral" (conveying no selective advantage or disadvantage) at a given time. However, the selective value of these genetic materials is not fixed eternally. What is advantageous under one set of environmental conditions may be disadvantageous under other circumstances.

In addition, human populations may respond to similar environmental stresses through unlike genetic changes. Malaria has also been endemic in the Mediterranean basin over many centuries, but the sickle hemoglobin allele is either absent or found at fairly low frequencies among populations now living in these areas. In Italy, Sardinia, Greece, and other countries in or near the shores of the Mediterranean, another variant allele involving hemoglobin synthesis, beta-thalassemia, is found in frequencies which show a striking correlation with the incidence of falciparum malaria. In Sardinia, for example, populations living in villages at lower altitudes, where malaria was endemic until only a few decades ago, have a higher frequency of individuals heterozygous for this allele than do populations living in villages found at altitudes above 400 meters. Occasional exceptions to this distributional pattern have been traced to the relatively recent settlement of some villages at lower altitudes by migrants from areas in which endemic malaria was not present. In Italy, the incidence of the beta-thalassemia allele also parallels the former distribution of endemic malaria, but in Thailand, where the thalassemia trait is also present, the correlation of trait and malaria incidence is not so clearcut.

In homozygous combination, the allele for thalassemia produces a debilitating anemia, known as thalassemia major or Cooley's anemia which, if untreated, frequently results in death during infancy or childhood. Yet the frequency of heterozygotes in certain Sardinian villages reaches as high as 38 percent. The medico-genetic picture is somewhat obscured by the existence of several variants of the thalassemia allele (both an alpha and beta form are recognized), and, in some populations, by the simultaneous presence of another trait, glucose-6-phosphate dehydrogenase deficiency (G-6-PD deficiency), which has also been correlated in some areas with the incidence of endemic malaria. It has been argued that the red blood cells of heterozygote carriers for any one of several of these hemoglobin variants do not provide such suitable conditions for the survival and multiplication of the parasite *Plasmodium falciparum* as do the red blood cells of normal *AA* homozygotes. But when counts have been made of the number of parasites in heterozygotes for some of the hemoglobin variants, the results have not always been consistent. Consequently, the presumed selective advantage of the heterozygote for hemoglobin variants other than sickle hemoglobin is based essentially on the correlation found between the presence of the variant allele at relatively high frequencies in populations subject to long periods of exposure to malarial environment.

Natural selection does not always result in the retention of diversity, but may instead take the form of *directional selection*—selection which increases the frequencies of some alleles and decreases the frequencies of others. At a given time, allele frequencies at a locus subject to this form of selection may have certain values, let us say $A = 0.90$ and $a = 0.10$; at a later point in time, these frequencies may have changed to the values $A = 0.01$ and $a = 0.99$. At this point, we can say that the earlier condition was one of *transient genetic polymorphism* in which two alleles at a locus were present at frequencies higher than can be explained as the result of mutation alone; but this condition was transitory as the gene pool of the population moved from one adaptive norm to another. Perhaps the most famous case in the genetic literature involving directional selection comes from studies in England of industrial melanism in the moth, *Biston betularia.*

Only a light grey form of this moth was reported in England until the middle of the nineteenth century when occasional black variants began to be observed. By 1895 nearly 98 percent of the moths studied in Manchester were of the black variety, and this dark variant became the dominant form throughout the industrialized areas of the country. It was conjectured that the darker variant was less likely to be spotted and eaten by predator birds on trees darkened by industrial pollutants—while in the past, and even today in non-industrialized parts of the country, the lighter moth had enjoyed a comparable advantage on the light-colored trunks of undarkened trees. Kettlewell (1965) tested this interpretation by releasing both forms of moths in two wooded areas—the first containing trees free of soot and overgrown with light-colored lichens, and the second area wooded by trees blackened with soot. Later, he recaptured the remaining live moths in the two areas and found a higher percentage of the light-colored moths remaining in the first wooded area, but relatively more dark variants in the second area.

In a number of other studies, insecticide-resistant forms of various insect species have been found to be the prevalent variety in populations exposed to such insecticides as DDT, dieldrin, malathione, and nicotine sulfate. Presumably, rare mutant forms of these insects may have appeared in populations prior to the spread of these insecticides, but in the past these mutants must have composed only a minute proportion of each population. As insecticides received widespread application, these rare mutants became the dominant form; as noted above, this directional change has hampered attempts to eradicate diseases such as malaria which involve transmission by an insect vector to human populations. The effects of directional selection in favor of antibiotic-resistant bacterial forms, as described earlier, poses an additional and potent threat to the health of modern human populations.

Natural selection in human populations is necessarily based on *ex post facto* studies; that is to say, selection is measured *after* differential reproduction has taken place. Additionally, because we cannot ethically subject human populations to many forms of experimental studies, verifications of hypotheses are not so readily at hand as are the kinds of confirming studies carried out on moths, bacteria, mosquitos, or other forms of animal life. Moreover, because the length of a human generation parallels the professional lifespan of an investigator, it is seldom possible to study a transient genetic polymorphism long enough to record an appreciable change in allele frequencies of the magnitude required to confirm the directional nature of the change. Nonetheless, it has been suggested that directional selection may now be affecting the distribution of hemoglobin *C*, another hemoglobin variant found in some areas of western Africa where endemic malaria exists. Individuals who are homozygous for this allele ($Hb^C Hb^C$) develop a milder form of anemia than that found in homozygous carriers of the sickle hemoglobin allele, while heterozygotes, $Hb^A Hb^C$, appear to have an increased resistance to complications from malarial infestation. In populations where both abnormal hemoglobin alleles coexist, heterozygotes with the genotype $Hb^S Hb^C$ suffer from a serious form of anemia which is, however, somewhat less severe than the anemia found in individuals homozygous for the sickle allele. It seems likely that the allele for hemoglobin *C* may be replacing the hemoglobin *S* allele in West African populations where both alleles are present.

A third form of natural selection, *stabilizing selection,* operates to maintain existing levels and kinds of polymorphisms, through the removal of harmful variants from a population which has attained a high degree of adaptive fitness in a limited range of environments. Although the effect of stabilizing selection is to maintain the status quo of the adapted population, this need not imply selection in favor of a single "perfect" genotype. Rather, this form of selection favors that range of genotypes (and the resultant phenotypes) which enables their carriers to survive and reproduce in the occupied environment. One often-cited example of stabilizing selection in human populations involves the range of optimal birth weights of newborn children.

Until very recently in human history, the pelvic delivery of a very large infant must have been attended by high risks to the life of both mother and child, while surgical intervention through Caesarian section all too frequently resulted in the death of the mother and, subsequently, to an increased chance of death in the child if it survived the surgical procedure of delivery. On the other hand, low birth weight infants, particularly immaturely developed or premature infants, were also at high risk. Selection undoubtedly operated in the past with great intensity against either very low or very high birth weights. Numerous recent studies, many summarized by Van Valin and Mellin (1967), have confirmed that ongoing selection for optimal birth weights can be demonstrated today in studies of *fetal deaths* (deaths before or during delivery) and in the mortality experience of *neonates* (infants from birth to age 28 days). These studies indicate that the probability of death increases at about an exponential rate with deviation from the optimal birth weight—which, in a Manhattan hospital series, was 3.62 kg for males and 3.84 kg for females. While the optimal mean birth weight seems to vary slightly from one population to another, selection against both extremes appears to characterize all human populations.

Whichever form of natural selection is operating, it is often useful to examine the force of selection and its consequences in more detail through the use of simple mathematical models. In reality, the value of s, the *coefficient of selection,* a measure of the intensity of selection operating against specific genotypes, is usually very small; but the basic principles of a mathematical description can be seen by looking at a hypothetical case involving values which are more readily manipulated. Suppose in a certain population that toe length was a trait involving codominant inheritance, so that long toes resulted from the genotype *TT,* short toes were due to the homozygous recessive genotype, *tt,* and medium length toes were found in individuals with the genotype *Tt.* Let us also stipulate that this population is living on a small island; that the allele frequencies for this trait were $T = 0.50$ and $t = 0.50$; and that the genotype distributions, generation after generation, had remained at approximately $TT = 0.25$, $Tt = 0.50$, and $tt = 0.25$.

One day, the island on which our hypothetical population lived was inundated by an immense tidal wave. About half of the 25 adults with long toes were drowned when their toes were caught in the debris carried along by the wave, while all the 25 adults with short toes drowned when they were unable to swim long enough to stay alive until the waters covering the island subsided. Fortunately, all 50 adults with toes of medium length survived, since their feet were neither caught in debris nor were they unable to remain afloat until all the wave had passed over the island. The value of s for the three genotypes would have been: $TT = 0.50$, $Tt = 0.0$, and $tt = 1.00$, these figures deriving from the mortality experiences of each genotype category. In a complementary expression, we could also describe the *Darwinian fitness,* or *W,* the success, of each genotype category as $TT = 0.50$, $Tt = 1.00$, and $tt = 0.0$. This can also be illustrated in diagrammatic form, as in Table 9-1.

Suppose now that the survivors of this catastrophe mate at random and that no evolutionary forces are operating, the genotype distributions of the F_1 population can be predicted from the Hardy-Weinberg formula as approximating: $TT = p^2 = (0.605)^2 = 0.37$; $Tt = 2\,pq = 2\,(0.605)$

TABLE 9-1 Natural Selection in a Hypothetical Population

	TT	*Tt*	*tt*	Σ
Preselection Period				
Allele frequencies:				
T = 0.50, t = 0.50				
Genotype numbers	25	50	25	100
Genotype frequencies	0.25	0.50	0.25	1.00
Selection				
s	0.50	0.0	1.00	
W	0.50	1.00	0.0	
Post Selection Period				
Genotype numbers	13	50	0	63
Genotype frequencies	0.21	0.79	0.0	1.00
Allele frequencies:				
T = 0.21 + 0.395 = 0.605				
t = 0.0 + 0.395 = 0.395				

(0.395) = 0.48; and $tt = q^2 = (0.395)^2 = 0.15$. Subsequent generations could be expected to approximate these genotype distributions so long as the necessary conditions of genetic equilibrium are maintained.

Although the foregoing example has focused on mortality differentials among genotypes, differential reproductive success is really the key to natural selection. Since members of the hypothetical population belonging to different genotypes died *before* reproducing, the gene pool of the F_1 generation was radically altered. But differentials in contributions to the gene pool of succeeding generations do not depend only on mortality differences of different genotypes. Differences in reproductive performance in the absence of mortality differentials by genotype can have the same effect. The following example, drawn from actual studies, demonstrates how selection can operate in part, at least, through differential reproduction of a heterozygote genotype.

Cystic fibrosis, briefly mentioned earlier, is a genetic disease characterized by the failure of secretory glands to produce mucus and necessary digestive substances. Even with advanced medical techniques and treatment, few afflicted children (who have the homozygous recessive genotype, *cc*) survive to reach even the early reproductive years. Yet the estimated incidence of the disease in several Euroamerican populations is close to 4.0 per 10,000, and the allele frequency is 0.016, or nearly 0.02. The maintenance of this allele at so high a frequency–much higher than could be maintained by mutation alone–constitutes a balanced genetic polymorphism. If the heterozygote carrier of the allele had only a small selective advantage over the homozygote for the *C* allele, the recessive allele could be maintained at the observed frequencies.

In the course of several investigations, researchers were unable to find evidence that heterozygotes had enhanced resistance for other diseases, which might have explained the hypothesized selective advantage of individuals with the *Cc* genotype. However, a significant difference between heterozygotes for cystic fibrosis and *CC* homozygotes was identified in the average size of the sibships for each genotype.

Knudson, Wayne, and Hallett (1967) compared the sibship size of parents of affected children and of a control group of parents who were mostly neighbors and friends of the study group, as well as the mean number of liveborn offspring of both groups. While the mean number of liveborn offspring of control families was 3.43, that for the study group was 4.34. Even when the values are calculated for mean number of offspring surviving to nine months of age for the two groups, a comparison shows a significant difference, with study group parents having a mean number of 4.21 children surviving to this age compared to a mean number of 3.41 for the control group. Using the simple model developed in the previous example, it is possible to test the level of s which might account for this polymorphism.

If the c allele frequency has a value of about 0.02, then C must have a value of about 0.98. Following the format used earlier, a tentative calculation can be made with the value of s set at 2 percent (see Table 9-2).

In fact, as Hallett and his co-workers show, a lower value for s, of the order of 0.001, operating over 23 generations, could explain the presently observed frequency of cystic fibrosis. The cause of the differential in reproductive performance by *CF* families is, of course, not known, but the application of a mathematical model for examining the data can provide a useful technique for evaluating the intensity of selection which must be operating on this locus. The reader may find it useful to apply this technique to analyze some of the data from various studies of natural selection in human populations presented in the remainder of this chapter.

Accounting for the vast number of known genetic polymorphisms found in human populations is a continuing challenge to the energies and talents of many investigators, and considerable time and effort has been dedicated to the study of one of the best documented polymor-

TABLE 9-2 A Tentative Model of Selection: Cystic Fibrosis

Preselection	*CC*	*Cc*	*cc*	
Allele frequencies,				
C = 0.98, c = 0.02				
Genotype numbers	9604	392	4	10,000
Genotype frequencies	0.9604	0.0392	0.0004	1.000
Selection				
s =	0.02	0.0	1.00	
W	0.98	1.00	0.0	
Postselection				
Genotype numbers	9412	392	0	9,804
Genotype frequencies	.9600	0.0400	0.0	1.000
Allele frequencies:				
C = 0.96 + 0.02 = 0.98, c = 0.0 + 0.02 = 0.02				

phisms, that of the *ABO(H)* blood groups. Some years ago, Alice Brues (1954) pointed out the limited range of polymorphisms at this locus in the human species and suggested that selective forces must be operating on this system. Later (1963), she applied computer simulation techniques to develop a model that achieved a close approximation to observed world frequencies by incorporating certain assumptions about the relative selection fitness of different genotypes at this locus. Over the years, many studies have attempted to identify the nature of the selective pressures which might be affecting the frequency of the alleles at this locus; one source of possible selective agents, disease, was immediately suspect.

Studies based on medical records compared healthy members of control groups, classified by *ABO(H)* blood group, with individuals, also categorized as to blood type, but who suffered from such medical disorders as duodenal or gastric ulcers, cancer, cirrhosis of the liver, etc. In a series of such studies, statistically significant differences were found in the incidence of various medical conditions in people belonging to each of the several major *ABO(H)* blood group phenotype classes. Thus, the incidence of duodenal and gastric ulcers was significantly higher among *O* blood group persons than in individuals with *A*, *B* or *AB* blood type. However, it is not clear how selection could be effectively removing the alleles implicated in these disorders from the same gene pool since these conditions are more likely to affect individuals past reproductive age who have already contributed to the gene pool of the succeeding generation.

In the case of associations between *ABO(H)* blood groups and certain infectious diseases which can affect individuals of any age group, the operation of natural selection can more reasonably be suspected. For example, the virus that causes smallpox, *Variola,* has immunological characteristics similar to those of the antigen found in persons with *A* blood type, and it has been argued that the antibody mechanisms of *A*-type persons may be delayed in action when infected with *Variola* virus. In contrast, the antibody systems of *O* or *B* type individuals, who already have circulating anti-*A* antibodies in their blood streams, more quickly recognize the foreign substance and react defensively to it. It could be predicted on the basis of this reasoning that mortality from smallpox and severity of symptoms would be greater in individuals with either *A* or *AB* blood type. Vogel and Chakravartti (1966) examined survivors of smallpox epidemics in India in 1964, 1965, and 1966, as well as a number of patients suffering from the disease in 1965 and 1966, classified by blood type. Survivors showed a higher frequency of *B* and *O* blood groups, and both mortality and symptom severity were higher in patients with *A* and *AB* blood types. Their findings are also consistent with the observed association between areas in which smallpox has been endemic in the past and the presence of populations with relatively low frequencies of the *A* allele. It has also been suggested that similarities in immunological properties between *Pasturella pestis,* the bacterium which causes plague, and the antigen (*H*) present in individuals with *O* blood type may result in differential mortality from plague in *O* blood type persons.

Although plague and smallpox no longer constitute serious worldwide health hazards, until quite recently each disease represented a major scourge to human populations. As such, these diseases may have played a vital role in the creation and maintenance of human genetic polymorphisms in the past. But another infectious disease, influenza, continues to affect human populations, and one form of the virus, influenza virus A_2, has a significantly higher incidence in *O* blood group individuals than in *A* type persons. Interestingly, very few associations have been found between *B* blood group and various medical disorders or diseases. Since this blood type is relatively low in frequency in most populations and many hundreds of thousands of cases need to be examined in order to detect significant associations where the intensity of selection is low, this is not too surprising. In fact, problems in the gathering of data and the statistical techniques used in analyses have led some investigators to question if any of these statistical associations provides adequate evidence for the role of disease selection on the *ABO(H)* polymorphisms.

Another line of investigation into the subject of *ABO(H)* polymorphisms is concerned with immunological interactions between mother and fetus. In the earlier discussion of gene flow (Chapter 7), the basic features of one maternal-fetal incompatibility interaction, that involving the *Rh* locus, were described. This interaction, as described in somewhat simplified form, requires the sensitization of an *Rh*-negative (*dd*) mother to the *Rh* positive (*D*-) antigen before she begins to produce anti-*Rh* positive (anti-*D*) antibodies. This may occur as a result of previous blood transfusions to the mother with blood of the incompatible *Rh*-type, or when the *D* antigen from the child is introduced into the mother's blood stream, either across the placental barrier during pregnancy or at the time of delivery. In the case of *ABO(H)* blood groups, the mother's blood already contains antibodies against the incompatible blood type of a fetus. The blood of an *O* type mother contains both anti-*A* and anti-*B* antibodies, for example, so that if sufficient levels of these antibodies cross the placental barrier, an *ABO*-incompatible fetus is at risk for erythroblastosis fetalis. Although rare, this condition has been observed in *A* or *B* children born to mothers with *O* blood type, and in other cases, such as *A* children with *B* mothers or *B* children with *A* mothers, where the mother's blood carries antibodies against antigens present in the blood of the fetus. Additionally, some studies have found a deficiency of births of children who would be expected to have an incompatible blood type to that of the mother, as in marriages of *O* blood type women to men with *A* or *B* blood type, so that it is possible that *ABO(H)* incompatibility maternal-fetal interactions may, at least in some cases, operate well before pregnancy can be detected or before the child can develop erythroblastosis fetalis.

Interestingly, the *ABO(H)* maternal-fetal incompatibility interaction may have a beneficial effect on the maintenance of the *Rh* polymorphism. If, for example, a woman with the blood type *B* and *Rh*-negative (genotype *BOdd*) is carrying an *A, Rh*-positive blood type child with the genotype *AODd,* the incompatibility at the *ABO(H)* locus may result in a spontaneous abortion before there is any sensitization of the mother's blood to the *Rh*-positive antigen produced by the fetus. A subsequent pregnancy, then, would not be at risk for *Rh*-incompatibility interaction, since the mother's blood would not yet have formed anti-*Rh* positive (anti-*D*) antibody. Thus, while it is usually necessary to describe natural selection as operating at a single locus, in fact interactions at several loci may actually be affecting polymorphisms at one or more genetic loci.

Another kind of complex interaction involving selection at the *ABO(H)* locus has been identified by Corinne Wood (1974, 1975). In this case, selection operates through the feeding preferences of *Anopheles gambiae,* the mosquito which transmits falciparum malaria. Originally, Dr. Wood's research was concerned with the possibility that mosquito feeding preferences might be affected by the skin color differences of human victims. She had volunteers, classified by skin color, expose their arms within an enclosed box to the ministrations of hungry mosquitos placed in the container. As part of the physical examination of the volunteers, each subject's blood was also tested and blood-typing tests were performed. As sometimes happens in scientific research, the hypothesis being tested—that mosquitos would feed differentially, dependent on the skin color of available targets—could not be proven; but when the data collected in these studies were examined to see if any other factors might be affecting the observed results, it was found that *A. gambiae* showed a marked preference for biting subjects with *O* blood type! Since mosquitos transmit several diseases affecting humans, the insects' feeding preferences, in this case for individuals with *O* blood type, may be a potent selective agency.

But how can a mosquito detect the blood type of an intended victim? It now appears likely that the answer may lie, in part, in the genotype of the individual at another locus, that for secretor status. In the presence of the dominant allele at this locus, *Se,* the antigens at the *ABO(H)* locus are expressed in various bodily secretions, such as saliva and sweat. Moreover, the molecular structures of the *A, B,* and *H* antigens differ in terms of sugars attached at specific positions to the carbohydrate chains of the precursor substance from which the antigens are formed. Further research will be required to elicit the precise mechanisms, but it seems that *A. gambiae* is able to detect slight differences in the antigenic characteristics present in bodily fluids, such as sweat, of potential victims.

Since much of the evidence on selection at this locus in human populations suggests that *A* and *B* antigens are subject to selective removal, as in incompatibility interactions and in a number of studies of *ABO(H)* blood type and disease correlations, Dr. Wood's investigations offer an example of countervailing selection against *O* blood type. It also appears that the presence of this allele at very high frequencies in native populations of South and Middle America provides independent evidence which can help to resolve the long-standing question about the antiquity of malaria in the New World. Had malaria been present among native Americans for thousands of years, then selection should have favored any *A* or *B* alleles in the population. The high frequency of the *O* allele in these populations argues for the recent introduction of the disease into the New World.

Future generations may look back at this stage of our understanding of natural selection affecting human populations with sympathy at our simplistic notions. Much of natural selection must surely involve complex, often subtle, kinds of interactions which we are only beginning now to detect. The foundations, at the crude level, have been established by examining relatively simple relationships between a single genetic locus and a specific environmental factor; but, as we increasingly become aware, interacting loci are affected by environmental factors which may be mediated through other organisms or via secondary levels of environmental influences. As yet, we can provide few satisfying hypotheses to explain population differences or polymorphisms involving such polygenic traits as stature, except to recognize that external factors may play an important role in many of these expressions of human variation. Most of our understanding of natural selection in human populations today has come from studies of blood group antigens and of hemoglobin variants, but it would be misleading to conclude this chapter without reference to studies of some of the other traits in which humans vary and which have also been subject to selective pressures.

One of the polymorphisms which has been studied at some length by anthropologists involves the ability to taste a certain artificially produced chemical, phenylthiocarbamide (PTC). The ability to detect this bitter-tasting chemical is controlled by the presence of a single allele, *T*, while non-tasters are homozygous for the recessive allele at this locus. Since the test is easily performed by having subjects taste a small paper strip impregnated with the chemical, the frequency of the phenotype has been determined for a number of populations. Although there is some degree of variation in expression, most subjects can be readily classified as tasters or non-tasters of this substance.

The frequency of tasters is very high (90 percent or more) in American Blacks, Chinese, and Peruvian Indians, but drops to lower values in many western European and Euroamerican groups. Indeed, because the trait seemed to have no selective value, differences in allele frequencies were sometimes used in earlier studies to estimate rates of admixture in mixed populations descended from ancestral groups which varied in frequencies of alleles at this locus. In the light of modern understanding, the maintenance of both alleles at the frequencies observed in various populations clearly constitutes a balanced polymorphism which suggests that selection must be operating.

As Greene (1974) has pointed out, this chemical has a goitrogenic effect—that is, it interferes with thyroid gland functioning. A number of compounds related to PTC are also found in such vegetables as brussel sprouts, kale, cabbage and other plants belonging to the genus *Brassica*, and tasters for PTC identify these compounds in such vegetables as bitter in taste. Thus, in areas where the diet is low in iodine, individuals who avoid foods which interfere with thyroid functioning because the foods are perceived as bitter in taste would avoid hypothyroidism, a condition that can be detrimental to physical and neurological development and which has been shown to be associated with reduced fertility. The selective advantage of tasters in low-iodine areas may, on the other hand, be balanced by other selective pressures operating against them. For example, various studies have found a higher prevalence of tasters among patients suffering from tuberculosis and in women with cancer of the ovaries, uterus and breast. It seems likely that selection, operating in some areas to the advantage of tasters, has been balanced by the actions of other selective agents which favor the non-taster phenotype. The result has been to maintain this polymorphism within the species, while populations in different locales have attained different allele frequencies at the locus corresponding to the specific stresses to which they have been exposed.

Among the polygenic traits which vary in frequency within and among populations, perhaps none has received more attention than skin pigmentation. In common with other polygenic traits, skin color phenotypes vary along a continuous scale of gradations rather than forming a limited number of discrete categories. With the advent of the reflectance spectrophotometer, an instrument by which skin pigmentation can be objectively measured, comparative results become readily available from studies in a large number of populations. It now seems clear that skin color involves five or more pairs of additively acting alleles, and that populations differ in the characteristic ranges of skin color present in members of different groups.

Skin color depends mainly on two factors—the amount of melanin present in the epidermis, and the blood supply found in small vessels in the dermis layer of the skin. Melanin is produced by specialized cells, melanocytes, which are found in the basal layer of the epidermis. While the number of melanocytes varies in different parts of the body, the total number of melanocytes does not vary significantly from one individual to the next, regardless of skin color. Rather, differences between individuals in skin color depend on the amount of melanin granules produced by the melanocytes. In persons with light skin color, only a few dark granules are present in the epidermis, so color of the skin is largely determined by the blood present in small blood vessels beneath the surface of the skin. Where numerous melanin granules are present, the contribution of the blood supply to skin color is masked by the dark material.

It had long been recognized that, in general, the distribution of skin color paralleled broad climatological zones. While all populations exhibit individual variation, darker skin color is more common in populations with a lengthy history of occupation in tropical areas of the world. In contrast, the frequency of light skin color is much higher in populations found in northern countries, such as England or the Scandinavian nations.

Exceptions to this distributional pattern can sometimes be explained in terms of historical events: American Blacks are relatively recent migrants to the New World; Polynesians in tropical regions of the Pacific are thought to have arrived in these areas only within the last millennium or two. However, American Indian groups show relatively little difference in skin color among tribes, though aboriginal Americans occupied a wide range of climatic zones. Also, dark-skinned Tasmanians inhabited a temperate zone. Thus, while exceptions certainly do exist, some of which are not readily explicable in the light of present knowledge, the general distributional pattern suggests a meaningful correlation between skin color and environmental factors.

In attempts to detect the evolutionary forces which might be responsible for this pattern, it was soon noted that ultraviolet radiation, which can be damaging to human tissues, was absorbed by melanin. Populations living in tropical regions of the earth, those parts of the world found between 23-1/2° north and south latitudes, are exposed to higher levels of solar radiation than are peoples living in temperate zones or in regions with frequent or extensive cloudy conditions. Arctic dwellers, exposed to clear skies and dust-free conditions during summer months and to reflection from snow and ice during much of the year, would be at somewhat greater risk from ultraviolet radiation than inhabitants of regions with more moderate climatic conditions. Accordingly, it has been argued that natural selection would have favored dark skin color in regions of maximum exposure to ultraviolet radiation. This argument has been augmented by the discovery that the incidence of skin cancer is lower among pigmented native inhabitants of sunny regions of the American Southwest, South Africa, and Australia than among more recent, light-skinned newcomers to these areas.

On the other hand, it has also been argued that light skin color offers a selective advantage to inhabitants of northern areas who must manufacture vitamin D from steroids present near the surface of the skin under the influence of ultraviolet radiation, and the amount of this radiation is limited in these areas. Failure to carry out this synthesis can result in rickets, with the resultant deformities and improper development of the bones associated with the disease. Were melanin present in larger amounts, the absorption of ultraviolet radiation would inhibit the manufacture of the vitamin, so light-skinned populations in areas where the amount of sunlight is limited would be at a selective advantage under such conditions.

While it is difficult to do more than hint at the broad range of strategies being employed to study natural selection in human populations in this brief survey, the topic is of central concern to our present knowledge of human variation. With rare exceptions, natural selection is claimed as an explanation of genetic polymorphisms in human populations when a plausible mechanism (disease, thyroid gland functioning, vitamin-D synthesis, etc.) can be reasonably claimed to account for an observed correlation between the incidence of a trait and its geographical distribution In fact, this is an eminently suitable procedure, since natural selection is a dynamic process which has not been ended, but augmented, by cultural evolution.

Medical advances make it possible for the adult diabetic to survive and reproduce through treatment of symptoms, but the allele or alleles responsible for the disease are thereby retained in the gene pool. Under differing environmental conditions, this genetic constitution may prove to have greater fitness than the genotypes of those whom today we would term "normals." The carrier for Tay-Sachs disease may, indeed, have had enhanced resistance to disease once endemic in European ghettos. As newly recognized mutant and variant forms of bacteria and viruses appear, heterozygotes for this and other genetic diseases may be found to have greater resistance to the sequelae of other communicable diseases. With the rise of modern, industrialized nations, medical knowledge and technology have advanced rapidly so that very few children die of typhoid fever and more of us survive to experience stress-related disorders, hypertension, and heart disease. Selection is an ongoing process in modern human populations which has the admirable function of providing the singular means by which our species can respond genetically to the inevitable certainty of change.

PROBLEM SET: NATURAL SELECTION

---

1. In a certain population of 1000 people, 16 percent of all individuals had N blood type, at least until the mysterious disease known as "galloping crud" was introduced by a visiting scientist. The disease spread rapidly through the population, and one-half of all people with M blood type died from it ($s = 0.50$, $W = 0.50$). Unfortunately, all those with N blood type survived the disease but succumbed to the treatment ($s = 1.00$, $W = 0.0$). The survivors of both the disease and the treatment resumed normal life within a few weeks; but when this population is examined in subsequent generations, we might expect to find the phenotype frequencies at this locus approximating $M =$ ________, $MN =$ ________, and $N =$ ________.
2. In certain parts of the United States, about 10 percent of the residents are American Blacks, and the allele frequency for sickel hemoglobin

among them is about 0.06. In one community with a total resident population of 100,000 people, plans are being made to develop a genetic screening program and adequate medical care for sicklemia victims. It would be economically feasible to develop the screening program if less than 1000 heterozygous carriers for this allele are likely to be present in the community. What is the expected number of individuals with sickle cell trait ________? the number of babies expected to be born with sicklemia ________?

3. If only half of the babies born with sicklemia in the community described above survive to reach reproductive maturity, what can the allele frequency for sickle hemoglobin be expected to be in the next generation of American Blacks in this area? *Si* = ________.

REFERENCES AND RECOMMENDED READINGS

ALLISON, A.C. 1954. Protection afforded by sickle-cell trait against subtertain malarial infection. *British Medical Journal* 1:290–294.

BRUES, A.M. 1963. Stochastic tests of selection in the ABO blood groups. *American Journal of Physical Anthropology* 21:287–299.

——. 1954. Selection and polymorphism in the ABO blood groups. *American Journal of Physical Anthropology* 12:559–597.

GREENE, L.S. 1974. Physical growth and development, neurological maturation and behavioral functioning in two Ecuadorian Andean communities in which goiter is endemic. II. PTC taste sensitivity and neurological maturation. *American Journal of Physical Anthropology* 41:139–152.

HARRIS, H. 1966. Enzyme polymorphisms in man. *Proceedings of the Royal Society,* B, 164:298–316.

——. 1959. Enzyme and protein polymorphisms in human populations. *British Medical Bulletin* 25:5–13.

KETTLEWELL, H.B.D. 1965. Insect survival and selection for pattern. *Science* 148:1290–1296.

KNUDSON, A.G., JR.; L. WAYNE, and W.Y. HALLETT 1967. On the selective advantage of cystic fibrosis heterozygotes. *American Journal of Human Genetics* 19:388–392.

LIVINGSTONE, FRANK B. 1958. Anthropological implications of sickle cell gene distribution in West Africa. *American Anthropologist* 60: 533-562.

McKUSICK, V.A. 1971. *Mendelian Inheritance in Man. Catalogs of autosomal dominant, autosomal recessive and X-linked phenotypes.* 3rd ed. Baltimore: Johns Hopkins Press.

POST, P.W. Comp. 1975. Anthropological aspects of pigmentation. *American Journal of Physical Anthropology* 43:381–443.

VAN VALEN, L., and G.W. MELLIN 1967. Selection in natural populations. 7. New York babies. *Annals of Human Genetics* 31:109–127.

VOGEL, F., and M.R. CHAKRAVARTTI 1966. ABO blood groups and smallpox in a rural population of West Bengal and Bihar (India). *Humangenetik* 3:166–180.

WOOD, C.S. 1975. New evidence for a late introduction of malaria into the New World. *Current Anthropology* 16:93–104.

——. 1974. Preferential feeding of Anopheles gambiae mosquitos on human subjects of blood group O: a relationship between the ABO polymorphism and malaria vectors. *Human Biology* 46:385–404.

Biocultural Interactions and Human Microevolution

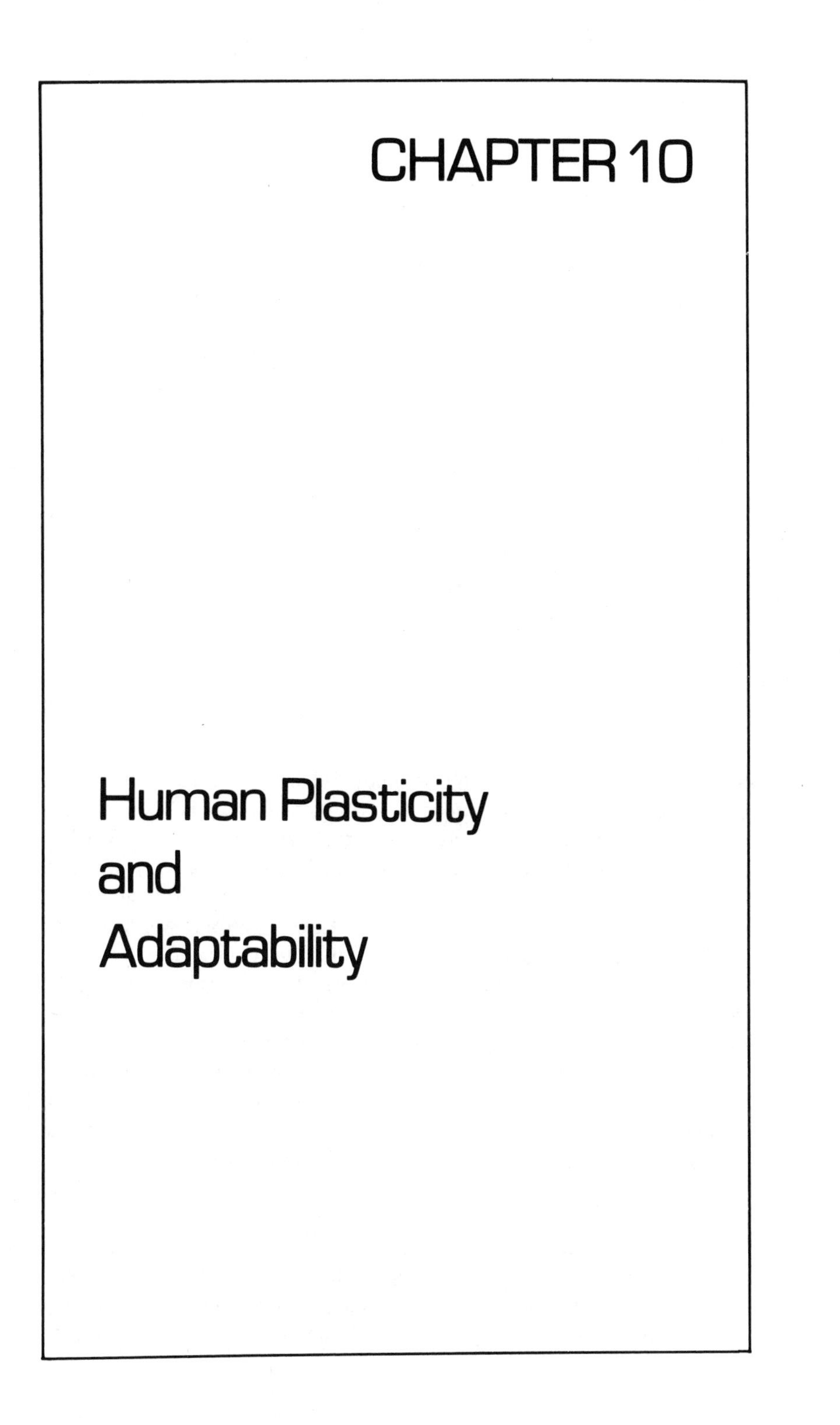

CHAPTER 10

Human Plasticity and Adaptability

Many of the differences which distinguish humans and human populations from one another involve genetic traits whose phenotypic expression is relatively invariant. Whether one is a Micronesian or an American black, a Hottentot or a Vietnamese, the same anti-*M* antisera will equally well detect the presence of the *M* antigen on the red blood cells of all individuals with the genotype *MM;* a single vial of anti-*A* antisera will detect the presence of the *A* antigen on the erythrocytes of each person tested whose genotype is *AA* or *AO.* But a number of genetic traits vary greatly in the range of phenotypic expression produced under varying circumstances; so, even identical, or monozygotic, twins may differ, for example, in adult height, in weight, in age at menarche, or in a number of other traits which have a genetic basis. As we have already noted, plasticity, the ability to respond to environmental stress by phenotypic modification, provides the organism with a mode of adjustment to environmental conditions which can complement the process of genetic adaptation through changes in allele frequencies.

The potential for modifications of this kind is, itself, the product of evolution. It is not difficult to imagine that a population endowed with genes which allow it to respond to cyclical fluctuations in food supply through changes in growth patterns would be more likely to survive periods of nutritional inadequacy than would a population lacking this ability. In the latter group, a dietary level inadequate to maintain the high nutritional demands of individuals whose genotypes are manifested in rapid growth rates and large adult body size must necessarily result in the genetic deaths of some proportion of the population. In a group in which nutritional stress is responded to by a decrease in rates of growth, adult body size may be limited as a result of chronic dietary restriction, but without causing loss of genetic materials from the gene pool of the population. Dietary restriction of sufficient severity and duration may produce irreversible changes in the phenotypes of members of such a group without any significant change in the gene pool; yet the offspring generation of such a modified parental population could express the full phenotypic range if raised under nutritionally adequate conditions.

The *adaptability* of a population consists of a variety of mechanisms by which it can respond to environmental pressures. Genetic adaptation is an adaptive strategy which results in changes in allele frequencies through the operation of natural selection, but these changes are relatively irreversible, at least over short periods of time. If malaria were eradicated tomorrow, it would take a number of generations for the sickle hemoglobin allele frequencies in West African populations to decrease to a level at which the allele would appear as a result of mutation alone; if malaria should subsequently reappear, it would require many more generations for the allele to resume its former frequencies. Moreover, genetic adaptation can lead to an evolutionary dead end as environments change, since the "best" genotype under one set of conditions may prove eminently disadvantageous under other circumstances. If directional selection has been effective, the survival of an isolated population that has undergone this form of selection would depend on the rare, chance appearance of a few mutant alleles which would be subject to the new selective pressures.

Traits which allow effective adjustments to different conditions through variable phenotypic expression provide another mode of biological response to environmental stress. Since such modifications in behavior, physiology, or morphology depend on inherited plasticity, it may seem arbitrary to distinguish them, as some biologists prefer, from genetic adaptation. However, to avoid confusion we will refer here to such changes, occurring within the lifetime of the individual, as *phenotypic adjustments* to environmental stresses. These adjustments, together with genetic adaptation, provide the biological mechanisms on which the survival of any species depends. In addition, cultural mechanisms provide the human species with distinctive modes of response to environmental pressures. Thus, as the following pages suggest, human adaptability comprehends a wider range of adaptive strategies—biological, biocultural, and cultural—than is found in any other biological organism.

The human species occupies a more widespread range of environments than any other Primate species, extending from the northern arctic regions to humid tropical forests and arid desert zones, living at altitudes from sea level to over 5,000 meters above sea level. The range of climatic conditions to which human populations are exposed today nearly corresponds with the total variation present on this planet; ancestors of living populations must have experienced an even wider variety of climatic conditions as weather and atmospheric conditions fluctuated during the course of human prehistory and evolution. The known distribution of early hominid fossils suggests that our ancestors were limited to tropical or subtropical areas of the Old World until at least a few hundred thousand years ago; so, most of human evolution must have taken place in the relatively benign climatic conditions of these zones. In a relatively short span of our total evolutionary history, then, humans have adjusted to the more severe cold conditions that must have prevailed in glacial and periglacial areas of prehistoric Europe and Asia, and to the extremes of heat and cold which human groups encountered as they moved into the more distant portions of the earth that are presently inhabited.

Every living organism possesses various regulatory mechanisms for maintaining a degree of constancy of the internal environment. These *homeostatic* mechanisms are vital in helping all warm-blooded animals, birds and mammals, to maintain the temperature of the vital organs—blood, heart, liver, intestines—within a limited range. The advantages of *endothermy* (the maintenance of a stable body temperature through internal regulation) are readily seen by comparison with animals whose temperatures fluctuate with those of the environment. Reptiles, for example, are unable to move rapidly on cool days, for their body temperature, heart beat, and metabolic rate are slowed during cool periods, and accelerate only as the radiant energy of the sun warms the body. Mammals and birds, in contrast, have a degree of environmental independence in this respect which confers a high degree of flexibility in behavior and activities.

Many of the chemical reactions in each cell are directed by enzymes, and these catalysts are temperature-sensitive. If the temperature falls below the optimum at which the reaction normally takes place, the metabolic reaction may proceed too slowly to produce the materials essential for vital processes to be carried out. Temperatures in excess of the optimum range for enzymatic activity may result in structural changes in the enzymes themselves, making them nonfunctional. In humans, the critical temperature range for survival is relatively narrow, between about 75° F. and 110° F. Thus, the maintenance of core body temperature within this limited range is an essential condition of life, and numerous sensitive homeostatic mechanisms are available to maintain this necessary degree of constancy.

Much of the fuel available from the food we ingest is released in the form of metabolic heat, produced by the body twenty-four hours a day, and additional heat may be produced by physical activity, exercise, or even shivering. Several kinds of responses occur almost immediately upon exposure to cold temperatures. First, the adrenal gland releases epinephrine and this hormone increases the metabolic rate. Reflex centers in the hypothalamus respond to stimuli from sensors in the skin and initiate impulses to the muscles so that shivering begins and more heat is generated. Blood vessels in the skin constrict (*vasoconstriction*) so that the loss of body heat through *radiation* (the flow of heat from a warmer to a cooler region) is reduced. In the hands, this initial vasoconstriction is followed in about five minutes by a dilation (*vasodilitation*) of the blood vessels. The pattern of alternating constriction and dilation which is established prevents the temperature of the tissues of the hands from dropping to levels at which frostbite damage can occur, while still minimizing the risk of loss of heat through radiation. At the same time as these changes are taking place, sweating also ceases.

While these responses appear in most subjects exposed to cold, the rates at which they take place and the efficiency of these responses vary among individuals and between members of different human populations. It now seems clear that both short-term and long-term adjustments which involve increasing tolerance to cold do occur, and a growing body of evidence suggests that more specialized biological responses to chronic conditions of cold stress have developed in some human populations. There is no question but that mechanisms of human adaptability in cold environments involve genetic adaptations or adjustments to this form of environmental stress.

Consider, for example, the ecological principle known as *Bergman's rule* which stipulates that body size of endotherms tends to increase as the temperature of the habitat decreases. This generalization has been shown to apply to a number of regional populations of mammalian species, and the reasoning behind the rule is entirely consistent with the need for heat conservation in cold climates. As body size increases, the ratio of skin area to volume decreases—so that heat production, which is proportionate to the mass of an animal, is maximized relative to the amount of skin exposed to the air. Heat loss, of course, is directly proportionate to the amount of exposed skin, so that we can safely predict that a large, preferably a spherical shape, body would provide one kind of solution to cold stress. *Allen's rule,* which states that extremities are reduced among populations of cold-dwelling endotherms, suggests also that regional populations of mammals could be expected to show a tendency for shorter legs and arms relative to trunk size in cold climates than would be found among related tropical species.

Among some (but not all) human populations, body size and shape conform to both the Bergman and Allen rules. Newman (1953, 1960) was able to demonstrate a significant negative correlation of weight to mean annual temperature for a large sample of native American Indians; Roberts (1973), based on a large sample of native populations throughout the world, confirmed that weight in males shows a significant negative correlation with temperature, and that natives of cold areas have larger chests and shorter arms than do native inhabitants of warmer regions. In Africa, however, where cold conditions are less severe than in the Northern hemisphere, increased body length appears to be positively correlated with increased temperature and humidity ranges. Similar body shapes, however, may be the product of different processes; it is usually not possible, for example, to determine whether small body size results from a specific genotype or nutritional insufficiency.

Attempts to study the effects of whole body cooling on human physiology suggest that regional populations may differ in cold-stress adjustments, at least in the kinds of responses which affect the extremities. One form of cold-stress adjustment, found in Europeans, Americans, American Blacks, Eskimos, and American Indians, among others, might be termed a "heat-production" response. To varying degrees and levels of efficiency, many members of these various populations respond to whole body cooling by an increase in metabolic rate. In comparison to Euroamericans, this metabolic response is slower and surface cooling greater in American Blacks, whereas among Eskimos the metabolic rate increase is more rapid and skin temperature is more elevated. An alternative mode of adjustment to cold stress, which might be termed a "heat-conservation" response, has been described by Scholander and others (1958) for native Australian desert dwellers. Among these desert aborigines exposed to freezing night temperatures, metabolic rate, as well as body and surface temperatures, drop while sleeping at sub-freezing temperatures. It is tempting to attribute these two different kinds of responses to genetic differences distinguishing Australian aborigines from other human populations, but subsequent studies (Hammel et al., 1959) have found that aborigines living under the more temperate conditions of Australia's coastal zones show a "heat-producing" response to cold stress, although the increase in metabolic rate among coastal natives is less extensive than observed in Euroamericans exposed to similar cold conditions.

In contrast, Steegman (1975) suggests that a number of differences among regional populations in response to extremity cooling may indicate genetically based differences in response to cold stress on hands or feet. Several studies have shown that a high percentage of American Blacks, in comparison to Euroamerican subjects, show no or delayed rewarming of hand tissues when exposed to immersion in cold water, while Euroamericans show a delayed rewarming effect in hand tissue in comparison to Eskimos. Comparisons made by Little and others (1971) of highland and lowland Peruvian Indians also provide strong evidence for a genetic base of response to extremity cooling, since both groups of Indians showed similar responses, although highland Indians, unlike those from coastal and lowland zones, were habitually exposed to a cold climate.

Obviously, it is difficult to unravel the effects of genotype and adjustment on the forms and efficiency of response to cold stress. Anecdotal evidence, no less than research, suggests that some individuals of all regional populations can develop better short-term responses to certain kinds of cold stress, and lifetime exposure may be required to develop the optimum response to cold stress which any specific genotype permits. Adult body build could depend on a long-term response to cold stress in which successful genotypes are those which, given existing dietary and disease conditions, undergo modifications in growth pattern and rate potentials to produce adults within the range of optimal body size and shape. Yet, marked differences among Eskimos, American Blacks, and Euroamericans subject to physiological cold stress tests are, at least, suggestive of significant genetic differences among human populations. Finally, cultural factors undoubtedly contribute to some of the observed differences among groups—the native Eskimo diet (high in fats and protein) no less than body build, enhances the ability to elevate metabolic rate and skin temperature. Euroamericans who have worked in the Arctic often report an increased desire for fat in the diet and an increasing tolerance for dietary fat as the length of stay in the region extends. Certainly, the adoption of Eskimo-type clothing has enabled Euroamericans to survive and work in the Arctic.

While further studies may help to specify the exact contribution of genetic and non-genetic factors, present data indicate that cold adaptability in human populations involves both genetic adaptation and phenotypic adjustment, modified by cultural variables. Despite the ability of Arctic dwellers, particularly the Eskimo, to create a protective micro-environment through such cultural forms as the heated igloo or the fur-lined parka, or to modify the internal environment through dietary practices, cold stress remains an imminent threat to survival in these climates. Our ancient ancestors in the Northern hemisphere, in Asia and Europe, were no less subject to these environmental pressures, so it is no wonder that physiological responses to cold stress, most notably to cold exposure of the extremities, are most effective in modern-day descendants of these ancient Europeans and Asians and less so among today's descendants of tropical dwelling Africans. Since it is most unlikely that natural selection is best effected in a widely distributed species by favoring an invariant or "perfect" genotype, human populations probably differ in respect to a range of genotypes conferring flexibility in expression in response to varying environmental conditions. Thus, humans may respond to cold stress with short-term and long-term physiological responses, which cannot now be readily associated with specific genotypes, or through more specialized physiological responses, which involve genetic differences. The response of desert-dwelling Australian aborigines to whole-body cooling through heat conservation, in contrast to the "heat-producing" response mechanisms seen in many other human groups, suggests that more than one mode of response, whether due to adaptation or adjustment, is possible. Moreover, cultural factors—behavioral patterns and extra-somatic modifications of the environment—which can enhance survival under cold-stress conditions are readily shared and exchanged by members of different populations who encounter similar environmental pressures.

The other temperature extreme to which humans must adjust if they are to survive is heat. Death is at least as likely to ensue if the body temperature exceeds about 110°F as if it falls below 75°F. If environmental temperature is lower than that of the body, which is normally around 98.6°F, heat from the body will transfer into the surroundings through radiation. When the external temperature is higher than that of the body, heat flows into the body through the same process; temperature-sensitive receptors in the skin are then stimulated; in response, messages from the hypothalamus produce expansion of the blood vessels (vasodilitation), a reduction in metabolic rate, increase in sweat gland activity, and reduced muscular activity. Through *convection* (the upward movement of hot air), the air close to the skin, now heated by the increased blood flow in extended blood vessels near the surface of the skin, is replaced by cooler air. Under ideal conditions, every gram of perspiration which is evaporated withdraws 540 calories of heat from the body, so one of the more effective responses to heat stress involves the production and evaporation of this fluid from the body. Although individuals vary in the total number of sweat glands, there is no evidence that there are significant differences between regional populations in this respect. This is true even though some of the populations examined for this trait live in conditions of high humidity where, of course, evaporative cooling is least efficient.

Human survival in hot and arid regions, where evaporative cooling of sweat would be most effective, could be enhanced if humans conformed to the logical implications of the Allen and Bergman rules applied to conditions of high-temperature environments. Thus, if body weight were low and body build linear, a maximum ratio of skin area to body mass would result and a larger surface area would be presented from which perspiration could evaporate. Various studies, such as Roberts's work mentioned earlier, have found a significant negative correlation between mean body weight and temperature, and a measure of limb length and increasing temperature. Schreider (1951) was able to show a decline in a body-weight to surface-area ratio with increase in temperature. Again, a few studies which have examined body composition indicate less thickness of the subcutaneous fat layer (which acts to reduce thermal conductance) among American Blacks than among Euroamericans. In addition, linearity is attained through delayed maturation in growth, and African children exhibit delayed skeletal growth patterns in comparison to European children of the same chronological age.

When exposed to experimental heat-stress conditions, there are usually no significant differences in responses among samples of subjects belonging to different ethnic or racial groups, when data are corrected for body weight and surface area. Occasional exceptions to this principle have involved small sample sizes (N = <10) or have failed to take account of such factors as recent dietary history, disease, age, etc. Most individuals from all populations seem able to adjust to some degree ("acclimatize") to increasing duration of exposure to heat stress conditions–gradually sweating becomes heavier and begins earlier, increases in body temperature and heart rate become less drastic. If, indeed, the human species evolved in tropical zones, and if early invasions of temperate and colder areas were made possible by behavioral and cultural adjustments, there is little reason to expect that important intraspecies differences would exist in the ability to adapt to heat stress.

Yet, human populations under similar temperature conditions do not uniformly conform to a single morphological mold. A number of West African populations do not fit the expected gradient of weight/surface ratio or limbs/weight ratio with temperature (Schreider, 1964), while tropical rain-forest dwellers, in general, provide a notable exception to the distributional pattern predicted by these ecological principles. In fact, temperature is only one of two factors defining total heat load, and the humidity conditions of tropical rain forest and island habitats are vastly different than those found in hot desert regimes. Humans are really subject to two qualitatively and quantitatively distinctive kinds of hot climates. In hot deserts with low humidity conditions, evaporative cooling is limited mainly by the body's ability to produce sweat. But the ability of the environment to vaporize sweat is limited by the amount of moisture in the air; so, evaporative cooling is always incomplete in hot humid conditions, and heat loss is governed primarily by the surface area of the individual. Some evaporative cooling does take place, but mechanisms to increase the amount of sweating would scarcely be advantageous. Rather, it appears, small size *and* body weight enable natives of the tropics to present a maximum surface area from which what evaporative cooling can occur does take place. Ideally, humans living under these conditions would have a minimum of adipose tissue, since medical studies clearly show an increased death rate from heat prostration in males of average or greater than average weight for age and stature. The slight body-build characteristic of many tropical populations closely parallels this model for humid heat adjustment.

It is important to note that humans must simultaneously adapt to a number of coinciding factors in the environment, and these multiple pressures may place conflicting demands on the body. For example, body fat reserves in the form of adipose tissue can provide a means of surviving periods of nutritional stress, but are antithetical with adjustment to prolonged heat stress. Among the Bushmen of the Kalihari desert, an unusual pattern of fat deposition results in *steatopygia,* the presence of large fat deposits in the buttocks area. This fat is not deposited throughout the body where it would act as insulation, and yet it provides the body with a ready form of stored energy to draw from in times of nutritional emergency. Could this be an adaptation to coinciding environmental pressures—heat stress and nutritional stress?

Life at high altitudes imposes a complex of environmental stresses —low oxygen pressure, low humidity, cold, and increased exposure to high solar radiation. Unlike heat or cold stress, *high altitude hypoxia* (inadequate oxygen in the lungs and body when the partial pressure of oxygen is low because of the reduced barometric pressure at high altitudes) can be alleviated only slightly, if at all, by behavioral or cultural adjustments. Thus, most of the responses to oxygen stress involve biological responses of both a short and long term nature.

The newcomer to high altitude is immediately faced with the problem of lower oxygen pressure in each breath of air drawn. At 4,000 m, only 60 percent as much oxygen is available as would be present when breathing at sea-level pressure (Stini, 1975). The immediate response, particularly if any physical exercise is called for, is one of an increased rate of breathing. In a few minutes, the increased loss of carbon dioxide which follows from this hyperventilation causes a shift in the acid/base balance of the blood to an alkaline condition. This, in turn, stimulates the production of epinephrine and other biochemicals; circulatory changes begin, involving reduced blood flow to the skin; and heart rate increases. In time, there is an increase in the production of certain red blood cell enzymes which facilitate the transfer of oxygen to the tissues, as well as an increase in the number of red blood cells released from the bone marrow into the blood stream. The increased number of red blood cells in circulation means that more hemoglobin is available, allowing for maximum utilization of whatever oxygen is available within the body. Other changes, such as those involving shifts in the excretion of certain ions by the kidneys (which helps to restore acid/base levels) also take place, as well as a number of biochemical changes which provide for other kinds of adjustments to oxygen stress.

In most adults, more long-term adjustments to high altitude follow as the length of sojourn in this environment extends over a period of years. The acid/base balance and respiration rate decline to approach previous sea-level values, although the viscosity, or "thickness," of the blood remains elevated as the number of circulating red blood cells continues to remain high. If the individual returns to sea level, the number of these circulating blood cells returns to pre-stress levels. In the native inhabitant of high altitude zones, certain characteristics related to high-altitude adjustments are commonly found. Large heart size relative to total body size is frequently observed, and probably results from the hypertrophy of an organ which must force viscous blood through large lungs containing a rich capillary bed. Commonly, too, native high-altitude dwellers have a characteristically large chest area, or thorax, encompassing larger lungs, and in which the sternum and ribs are enlarged, thus accommodating a greater bone marrow area in which red blood cells are manufactured.

Among these natives, body build is characteristically small, although such organs as the heart and lung are not reduced proportional to lower body weight. Studies of growth patterns in highland populations show that trunk growth is rapid in children while growth of arms and legs is slowed. The adult, then, has a large body mass/surface area ratio which, as we have already seen, minimizes heat loss, and this would be advantageous in cold climates. However, Stini (1972) has suggested that the diminished body size of high-altitude natives may be beneficial where chronic food shortage or episodic famine occurs, and high-altitude regions are characteristically areas of marginal agricultural productivity because of the cold arid climate, hilly terrain, and poor soils usually found in these regions. One of the frequent results of acute hypoxia is a large loss in body weight and a mobilization of fat stores. With prolonged exposure, protein catabolism occurs; animal studies show that mortality increases in small laboratory animals exposed to hypoxic conditions when fat is present, even in moderate amounts, in the diet. It would appear that the "Eskimo diet," high in fats and protein, which is beneficial in Arctic cold-stress responses, is inimical to high-altitude adjustments.

Mazess (1975) concludes that the relatively greater ability of highland natives to function in the high-altitude environment results from long-term exposure and reflects genetic plasticity rather than any unique genetic composition of highland populations. Attempts to show that genetic adaptation has taken place have focused on the purported low reproductive success of highland natives and, especially, of immigrants and newcomers to high altitude regions. Colonial reports indicated that Spanish women living in the Peruvian highlands were less fertile than their countrywomen living in lowland Peru, and an examination of recent census data for Peru seems to confirm the lower reproductive success of highland natives. Similarly, birth rates and infant mortality rates taken from census reports for mountainous areas of the United States showed apparently lower reproductive success among highland populations in this country. Grahn and Kratchman (1963), as noted in a previous chapter, showed that the higher neonatal death rates reported from mountain states do not reflect lower *fecundity* (the biological capacity for reproduction) or increased rates of genetic damage due to radiation, but are attributable to lower birth weights resulting from reduced fetal growth. Other variables distinguishing highland and lowland Peruvian Indians—dietary, medical, socioeconomic, demographic—may also account for the apparent lower fertility inferred from census data for highland Peru.

It would be surprising, given our present knowledge of human genetic variation, if populations did not differ in genetic composition in respect to some of the traits discussed here which enable humans to successfully inhabit very different environmental regimes. That the evidence for such differences is so scant suggests that no population totally lacks the necessary genetic plasticity to make adequate physiological, anatomical, and behavioral responses to a wide range of environmental pressures. Specific genetic differences, if these exist at all, are most likely to involve responses to cold stress, and this is entirely consistent with what we know of human evolutionary history. We evolved in the tropics, and the vast majority of hominid evolution has taken place in warm climates. As humans spread to temperate zones, clothing and the use of fire must have facilitated our occupation of areas subject to freezing, even periglacial, conditions. Some individuals, unable to make adjustments, surely died, and selection must have operated to maintain a minimal degree of plasticity within each population subject to comparable environmental pressures. Over a period of tens of thousands of years, specific adaptive genotypes may have become more prevalent in some populations than in others, and different kinds of adaptive responses (heat-conserving, heat-producing) may have been favored in different populations. The occupation of high-altitude zones, which must have taken place more recently in human history, has probably provided an inadequate length of time in which distinctive genetic differentiation could have taken place. Surely, the quest might better be directed, from an attempt to distinguish adjustments which reflect genetic adaptation to specific stresses from those which do not, to a search for greater knowledge of the physiological, biochemical, and growth processes which make human adaptability to complexes of environmental pressures possible at all.

REFERENCES AND RECOMMENDED READING

ADAMS, T., and B.G. COVINO 1958. Racial variations to a standardized cold stress. *Journal of Applied Physiology,* 12:9–12.

BAKER, P.T. 1971. Human biological diversity as an adaptive response to the environment. In *The Biological and Social Meaning of Race.* R.H. Osborne, ed. San Francisco: W.H. Freeman and Company, pp 25-56.

——. 1969. Human adaptation to high altitude. *Science,* 163:1149–1156.

GRAHN, D., and J. KRATCHMAN 1963. Variation in neonatal death rate and birth weight in the United States and possible relations to environmental radiation, geology and altitude. *American Journal of Human Genetics,* 15:329–352.

HAMMEL, H.T.; R.W. ELSNER; D.H. LeMESSURIER; H.T. ANDERSON; and F.A. MILAN 1959. Thermal and metabolic responses of Australian aborigines. *Journal of Applied Physiology,* 14:605–615.

LITTLE, M.A.; R.B. THOMAS; R.B. MAZESS; and P.T. BAKER 1971. Population differences and developmental changes in extremity temperature responses to cold among Andean Indians. *Human Biology,* 43:70–91.

MAZESS, R.B. 1975. Human adaptation to high altitude. In *Psysiological Anthropology.* A. Damon, ed. New York: Oxford University Press, pp. 167–209.

NEWMAN, M.T. 1960. Adaptations in the physique of American aborigines to nutritional factors. *Human Biology,* 32:288–313.

——. 1953. The application of ecological rules to the racial anthropology of the aboriginal New World. *American Anthropologist,* 55:311–325.

NEWMAN, R.T. 1975. Human adapatation to heat. In *Physiological Anthropology.* A. Damon, ed. New York: Oxford University Press, pp. 80–92.

ROBERTS, D.F. 1973. Climate and human variability. *Addison-Wesley Module in Anthropology,* No. 34. Reading, Mass.: Addison-Wesley.

SCHOLANDER, P.F.; H.T. HAMMEL; J.S. HART; D.H. LeMESSURIER; and J. STEEN 1958. Cold adaptation in Australian aborigines. *Journal of Applied Physiology,* 13:211–218.

SCHREIDER, E. 1964. Ecological rules, body-heat regulation and human evolution. *Evolution,* 18:1–9

——. 1951. Race, constitution, thermolyse. *Revue Scientifique Paris,* 89: 110–119.

STEEGMANN, A.T., Jr. 1975. Human adaptation to cold. In *Physiological Anthropology*. A Damon, ed. New York: Oxford University Press, pp. 130-166.

STINI, W.A. 1975. *Ecology and Human Adaptation.* Dubuque: Wm. C. Brown Company.

—— 1972. Reduced sexual dimorphism in upper arm muscle circumference associated with protein-deficient diet in a South American population. *American Journal of Physical Anthropology,* 36: 341-352.

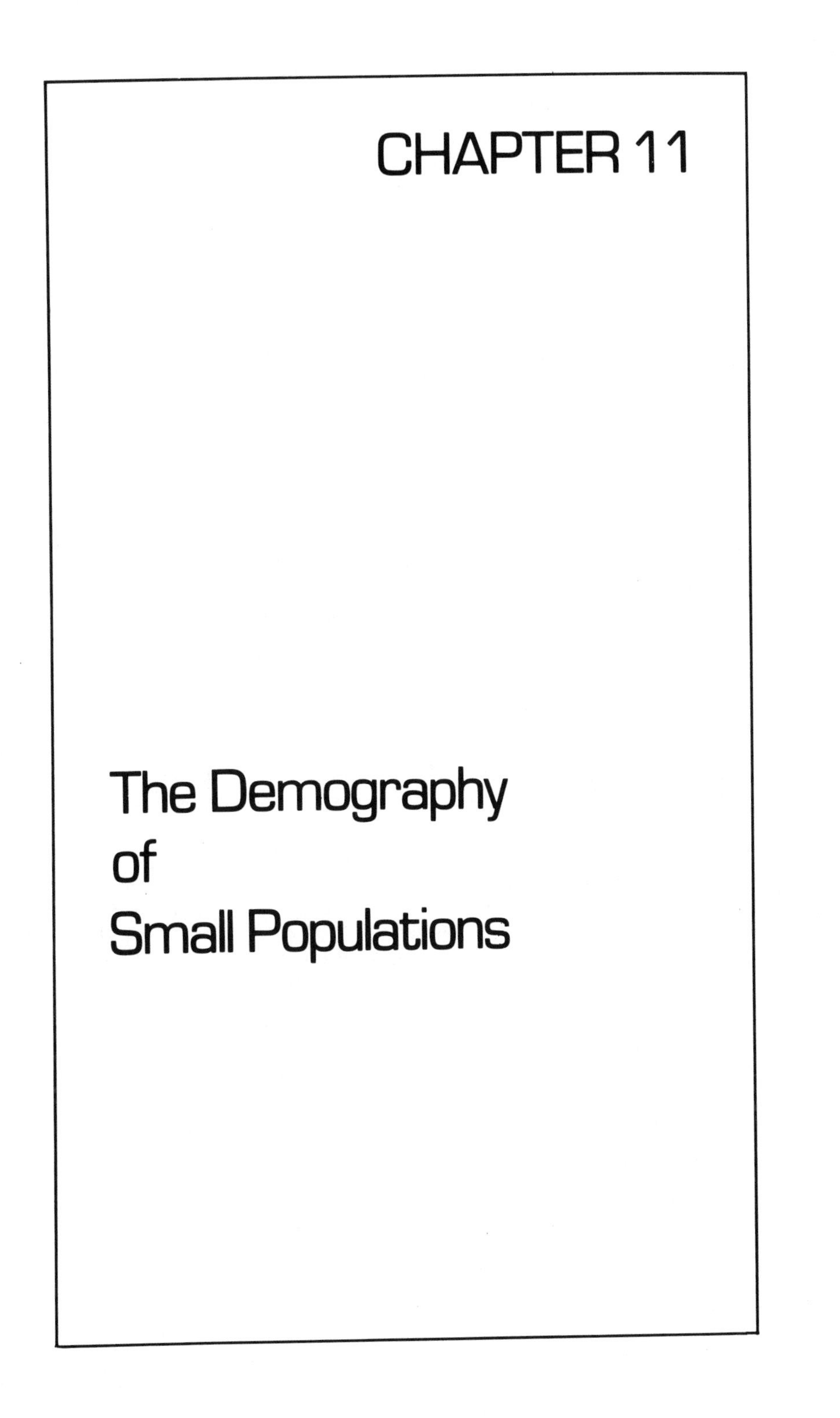

CHAPTER 11

The Demography of Small Populations

A modern student of human variation has succinctly described the tasks and goals of microevolutionary studies in the following words:

> We have stressed that the exact ways in which the interaction (of evolutionary mechanisms) takes place depends upon the environmental pressures, the sociocultural system, the population size and density, and the technological level. It is one task of human biologists to unravel these interactions in specific populations and to develop a set of generalized statements that can be applied to a wide range of situations. The task is far from easy or quick, but it is the one way in which microevolutionary process in man will finally be understood.[1]

In fact, for human populations the effects of evolutionary forces, or of departures from panmixia, cannot even be described apart from such characteristics as the size of a population, the distribution of its members in space, and the sex and age composition of the group. Gene flow, for example, cannot readily occur between two formerly distinct populations if one group is composed entirely of elderly men and women and the other

[1] Reprinted from F. Johnston. 1973. *Microevolution of Human Populations,* p. 122. By permission of Prentice-Hall.

has a more normal age and sex distribution. Natural selection, differential contribution of some genotypes to the gene pool of succeeding generations, can only be measured in terms of fertility and/or mortality differences, while the size of the effective breeding population directly specifies the conditions in which genetic drift operates. Even mutation is affected by population size and composition. In a numerically small population, a mutation rate of 1 per 100,000 gametes implies that a particular mutation has a low probability of appearing at all in any one generation, but could more probably be expected to occur in a population numbering many thousands. In a population containing a large proportion of aged persons, a rare, newly arisen gametic mutation may not be passed on at all if it occurs in the gametes of any of the numerous individuals past the age of reproductive activity. Obviously, if our task is to understand microevolutionary process, we have to examine underlying demographic factors, what Bogue (1969) has termed demographic processes—fertility, mortality, marriage, migration, and mobility—responsible for the observed population composition and structure.

In large-scale societies which maintain appropriate written records, vital statistics covering births, deaths, marriages, and migration are available from registers compiled by central agencies. Many countries and nations have made the reporting of these events compulsory, with failure to comply with these regulations subject to punishment. A number of countries have also conducted censuses of the entire population at more or less regular intervals for a number of years, and have developed centers for elaborate and sophisticated demographic analyses of such materials. Because the population size of nations and countries which collect and analyze these materials is often very large, random or temporary fluctuations in vital events have limited significance, and demographers have developed elegant techniques for handling erratic deviations, evaluating temporary perturbations, and detecting the effects of systematic errors in the data. Errors and shortcomings do occur in national censuses taken under even the best of circumstances: the 1960 United States census reported that over 1500 fourteen-year-old widowers were then living in the country, which seems an unlikely number of juvenile males to have experienced the death of a spouse. The Census Bureau has also had to institute special surveys to enumerate the members of certain ethnic minority groups in the United States. Nonetheless, vital statistics and census data from most of the industrialized countries of the world have become steadily more reliable over the years and are readily amenable to refined treatment and analysis.

Many of the societies which anthropologists study, however, are limited in size and often have, at best, inadequate systems of recording vital events. Many have only recently introduced census procedures, and must train enumerators and develop adequate procedures appropriate to local conditions. Among these populations even the age distribution is difficult to establish, since many persons may lack any proof of their date of birth. Acsadi and Nemeskeri (1970), for example, conclude that the high proportion of very elderly individuals reported for such countries as Iraq, Bolivia, Egypt, or the Fiji Islands is essentially an artifact of inaccurate birthdate information. McArthur (1968) reports a systematic error in digit preferences of Pacific island respondents to questions on date of birth. Countless other examples exist which demonstrate that various forms of systematic and nonsystematic error confound the accuracy of population counts in developing countries. In addition, random fluctuations in vital events have a potentially great effect on the structure and composition of small populations. As discussed earlier in connection with the subject of genetic drift, the effect of a random event can be magnified where total population size is small. Yet in small-scale societies, the genetic and evolutionary consequences of rare or random events, which are potentially enormous, are sometimes overlooked in the development of demographic models or of life-table models.

Despite these problems, anthropologists have increasingly turned to demographic measures of human adaptability. This is because the size, composition, and structure of all animal populations, including humans, are sensitive to biological, social, and environmental pressures. In order for a population of any species to occupy an area and utilize its resources successfully over a period of time, it must attain and sustain some degree of balance between the number of individuals in the group and the resources available to it. Some species produce a prodigious number of young, but relatively few of these will survive predation by other organisms or all of the other vicissitudes of life; so, relatively few individuals survive to reach reproductive maturity. In humans, usually only one offspring is produced at a time, and comparatively few offspring are produced by any one woman over the course of her entire reproductive lifespan. The ability of any human group over time and in a specific limited area to produce sufficient numbers of offspring, who survive to reproduce and replace themselves, depends on the balance which is maintained between the fertility and mortality experiences of each such group.

Much of human history was probably experienced by humans organized into small bands of individuals,, engaged in hunting and foraging of locally available resources. Although the precise manner in which these small groups controlled population size and distribution in space is not directly evident in fossil remains, the study of contemporary groups of hunters and gatherers can provide some clues to the reconstruction of the kinds of behavioral patterns which could have affected a long-term balance of population and environmental factors.

Ludwick Krzywicki (1934), a pioneer in the study of the vital statistics of native peoples, found that certain practices and conditions conducive to maintaining high fertility rates were found among such diverse nonagricultural peoples as Eskimos, Australian aborigines, and a number of American Indian tribes, among others. Unmarried adult females were rarely found and barren women were uncommon in these groups. Even where adult males were outnumbered by adult females, polygynous unions, in which a man had two or more wives simultaneously, were reportedly commonplace. The percentage of children in such groups was characteristically high, yet these populations remained essentially stationary in numbers. Krzywicki suggested that high rates of mortality, particularly among children, due to deaths from natural causes and from intentional infanticide, sufficed to maintain tribal populations of hunters and gatherers at a stationary level.

The Australian aborigines, who were hunters and gatherers until very recently, are thought to have numbered a few hundred thousand individuals, organized into slightly more than 500 tribes in the eighteenth century, at the time when Europeans first began to colonize that continent. As among a number of other small human groups dependent upon localized resources, the daily living group, or band, consisted of about twenty-five individuals. As Birdsell (1968) has shown, most nuclear families included between one and five children who survived to adulthood. An examination of differences in the sex ratio, the relative number of males to females, for different age groups reveals that systematic infanticide, preferentially directed against females, was an impotant factor in limiting population growth among these tribes. Similar findings among other tribal groups led Birdsell to contend that systematic infanticide, involving between 15 and 50 percent of all births, was a necessary procedure for child spacing throughout the entire period when humans followed a hunting and gathering way of life. Thus, among such groups, female fertility (actual reproductive performance) does not match maximum fecundity (the maximum biological potential for reproduction); and even further, the numbers of a group are maintained in part through intentional infanticide.

This is a far different picture of human history than that presented by those who portray human populations growing at a prodigious rate until local resources become decimated by human exploitation and the desperate survivors either succumb, themselves, to starvation or descend upon their hapless, but more well-circumstanced, neighbors. In fact, only in a very few known cases, where economically simple human groups have moved into previously unexploited areas, have populations doubled—or, in one case, tripled—in numbers in the span of one generation; and this rate of growth has been maintained for only brief periods of time (Birdsell 1957). Even in a mobile group, where a woman would be seriously disadvantaged if she had to carry and feed two small children at one time, she should still be able, theoretically, to successfully bear a child about once every three years, to produce about ten children in the course of her reproductive lifetime of about thirty years. Most women in economically simple societies do not produce this number of children; so, human fertility must be limited by a number of factors.

As discussed in an earlier chapter, Frisch has claimed[2] that menarche and the initiation of regular ovulation may be delayed in girls who do not attain a critical ratio of body fat to body mass; so, the period of adolescent sterility might be extended for girls in populations where nutritional standards and/or physical activity limit the accumulation of body fat deposits. It is also possible that irregular ovulation and the cessation of ovulation with menopause would occur earlier among such groups. Howell (1976), in fact, reports that menarche is delayed and the age of menopause lowered among the !Kung Bushman, a contemporary African pre-agricultural people. Breast feeding has, at best, an unreliable effect on ovulation, but probably contributes to some degree to reduced fertility; cultural restraints on sexual activity may more effectively reduce the risk of pregnancy. Nag (1962), for example, concludes from his study of sixty-one preindustrial societies that the removal of sexual tabus has been a significant factor in changing fertility patterns of developing countries.

Among the contemporary Hutterites of North America, who have ready access to advanced medical care and place a high value on fertility, the average number of children who will be born to women by the end of their reproductive lifespans is 10.4 (Total Fertility Rate). This value probably represents a maximum fertility performance for females belonging to a sedentary, agricultural society. Coital tabus are not reported for this group, and the opportunity for intercourse is maximized in a group which encourages stable marital unions. Excellent medical care optimizes the conditions for successful gestation and birth, while birth control is discountenanced and would be difficult to practice in this religious communal society. Given the far less advantageous circumstances under which early human groups must have lived, it is unlikely that many women would have given birth to more than half the number of children which Hutterite women produce during their years of reproductivity activity.

[2]Several studies (Johnston et al., 1971; Billewicz et al., 1976; Van't Hoft and Roede, 1977; and others) have seriously challenged this claim.

Thus, we can somewhat arbitrarily select a value of five as the average number of children that a woman living in a hunting and gathering society would give birth to. In fact, Howell reports a total fertility rate of five for !Kung women.

If all these children survived to become reproductively active adults, however, the population would necessarily increase in size over time. This level of fertility must be balanced by a compensatory level of mortality if numbers are to remain stationary in a localized population. We know that, in many contemporaneous groups of hunters and gatherers, bands are usually exogamous in nature and members must seek spouses from other groups. This condition ideally should result in an equal exchange of members between groups who maintain a marital exchange relationship with each other. But demographic factors often interfere with these ideal patterns: one band may have an excess of young unmarried women available for exchange with an adjacent group; another group may only have several older widows as possible wives for young males seeking a first marital union. Yengoyan (1968) has shown that the number of eligible mates, given prevailing marital restrictions, is a function of population size. In eight small Australian tribes which were divided into exogamous subsections, more than 40 percent of all marital unions were irregular; that is, they involved spouses who did not belong to the proper category of eligible mates. However, marital migration is not likely to have been a major form of size constraint in human populations, despite possible short-term changes in numbers produced as a result of occasional unequal exchanges of mates.

Kryzwicki, who contended that mortality in general was higher among primitive peoples than among Europeans, candidly admitted that he lacked suitable evidence to support this conclusion. Rather, he pointed to the higher crude death rates reported from various eastern European populations in earlier historic times to suggest that peoples who had not attained the standards of living found in these countries in the twentieth century must necessarily have more closely approximated the demographic characteristics of premodern European populations. Unfortunately, there are many problems in determining the mortality patterns of other societies, and there are cogent reasons for challenging the validity of extrapolating from the mortality experiences of preindustrial European populations to a reconstruction of mortality patterns in preagricultural societies.

Crude death rates—and crude birth rates, too, for that matter—represent a ratio between the number of events (births, deaths) to the total number of individuals in the population at the midyear point. This value is usually expressed as the number of events per 1000 persons alive at midyear:

$$\textit{Crude rate} = \frac{N_{(\text{births, deaths})}}{\text{Total N at mid-year}} \times 1000$$

This index takes no account of the age and sex distribution of the population: to take an extreme example, it can readily be seen that 100 deaths in a population of 1000 individuals aged 80 years or older (crude death rate = 100) is a very different situation from that which occurs when 100 deaths occur in a population composed of 1000 young adults and children (crude death rate = 100). Similarly, reports which provide only the average or median age at death, and this applies particularly to skeletal collections, are misleading. If, as has happened, skeletal samples include a large number of infants and young children, the resultant figure for median age at death is often so low as to make one wonder how enough people survived long enough to ensure the continuation of the population. Average values of this kind certainly contribute to the misconception that modern populations are facing a novel situation in supporting a respectable proportion of elderly persons. In fact, as Lee (1968) notes, in a total population of 466 !Kung Bushmen of the Dobe area, 46 individuals were age 60 or older, a figure which compares very well with the proportion of elderly in industrialized nations. Far more informative for understanding the ecology of living populations and the evolution of the human species are *age-specific vital rates*—the ratio between the number of events being studied and the numbers of individuals belonging to different age groups, and, thus, at greater or lesser "risk" of having such an event occur.

Needless to say, the collection of the necessary data to calculate these rates reliably is fraught with difficulty. In contemporary anthropological populations, where date of birth can rarely be firmly established, it is often difficult to determine the precise age and sex distribution of the entire population. Estimates of relative age may be collected and connected to specific known dates (e.g., "born after the Year of Falling Stars"), or age groups can be extended to cover such broad categories as "juvenile," "young adults," "old adults." Although quantitative data on age at death can be estimated from skeletal materials, the inevitable range of error in such estimates may be as great as the age range of the cohort to which the individual skeleton is assigned.

In many of these populations, the recent past has been a period of radical demographic change and the vital statistics which can be garnered do not reflect a stable pattern. Among most Pacific island populations, for example, population size had declined radically from the period of European colonization until after World War II. On the island of Yap, the documented population decline—from over 7000 natives resident in 1901 to less than 3000 natives in 1946—was reversed after the introduction of antibiotics in the postwar years. While the mean number of children produced by women born before 1922 was about two, women born after that time, who are now nearing the completion of their reproductive years, have had an average of more than four children each, presumably due to the control of venereal disease through antibiotic treatment. The Yapese are clearly beginning a period of rapid population growth, despite the fact that the mortality rate for children aged 0–4 years actually increased between 1946 and 1966 (Underwood, 1973).

In seeking to reconstruct the mortality patterns of prehistoric groups of hunters and gatherers, various investigators have been given pause by the problem of detecting the probable causes of death among our distant ancestors. The history of human population growth over the past few million years of hominid evolution is usually portrayed as a nearly straight line for most of the evolutionary period during which the numbers of hominids almost imperceptibly increased to reach perhaps a million. With the invention and spread of agriculture, population size began to increase greatly; and although there were undoubtedly periods of setbacks and episodic declines, the general trend in human population size has been toward a high, even exponential, rate of growth only since about A.D. 1100, until today we number about four billion.

If, as seems likely, our ancient ancestors were limited in numbers throughout most of the Pleistocene, the prevalent mortality pattern during that long period must have adequately supplemented whatever degree of family size control was being practiced. Although many contemporary populations of hunters and gatherers have already undergone rapid and extensive destabilization, certain characteristics common to this way of life probably can provide clues to reconstructing prehistoric conditions of disease and mortality. Dunn (1968), for example, argues that malnutrition and starvation occurred only rarely, since most hunter-gatherers exploited a diverse range of resources within the exploited area, and this must have been particularly true before humans invaded regions subject to seasonal food or water shortages. Indeed, recent studies among the !Kung Bushmen of the Kalahari Desert have upset a long-held view that hunters and gatherers live in a feast-or-famine world. Rather, as Lee has shown, adults of this group, working only about twelve to nineteen hours a week, are able to maintain the entire population in a state of apparent nutritional adequacy!

Many of the infectious diseases which have decimated large segments of human populations of the historical period, such as the pneumonic form of plague, smallpox, or even measles in some island groups, are transmitted by direct human contact (i.e., *contagious diseases*). Among small groups of hunters and gatherers living in relative isolation from one another, the necessary conditions for the successful maintenance and spread of the causative organisms of these maladies are minimal. The characteristically low population density and the mobility of hunting and gathering populations would also not have been conducive to the spread of such diseases as typhoid fever or hepatitis which can be transmitted indirectly through contamination of water sources. As discussed earlier, even those diseases which require an intermediate animal vector, such as falciparum malaria, may have had little impact on the size of preagricultural populations.

For certain other possible causes of death, such as accidents, trauma, or predation, Dunn notes great variations in frequency and occurrence among populations of contemporary hunters and gatherers. But "social mortality" (e.g., death due to cannibalism, infanticide, suicide) is said by Dunn to have played an important part in the population equation of hunting and gathering populations of the past and even those of today. Thus, while for some possible causes of mortality no generalization can apply to all groups, certain patterns of mortality common to hunting and gathering groups can be used to begin to interpret the course of human population history.

Throughout the Pleistocene, human populations seem to have established and maintained a long-term equilibrium between numbers—or, more properly, biomass—and environmental resources available through intensive exploitation by small bands of humans equipped with a limited technology for the hunting and gathering of localized resources. This relationship has been described by Birdsell as constituting a density-equilibrium system in which short-term oscillations in numbers are met by restorative changes in fertility and mortality. With the advent of agriculture and the sedentary way of life which often accompanies the cultivation of plants, the course of human population history was drastically altered. Densely settled villages provided suitable host populations for the maintenance and spread of pathogenic microorganisms; sedentary populations without adequate understanding of the prophylactic value of hygiene and sanitation polluted the environment; changes in the relationship of local human populations to other animal populations, both domesticated (cattle, sheep, goats) and undomesticated (rats, vermin), meant new or more intensive kinds of exposure to noncontagious infectious diseases. The protection of established communities and the defense of lands and structures that developed with the labor of generations all entailed new kinds of political relationships and the beginning of warfare.

While we cannot extrapolate from the !Kung to infer that all hunters and gatherers led the affluent life, it is clear that many agriculturalists work long hours to wrest a bare subsistence level of existence from the soil, and they are subject to periods of famine when rains do not appear at the right time or fail to come at all, when crop pests destroy a whole year's harvest, or when some other natural calamity interferes with the ability of humans to grow and store crops. In the absence of modern agricultural techniques and mechanized equipment, many forms of agricultural activities are labor-intensive. Productivity can be improved by expending additional labor on the clearing of lands, the sowing and harvesting of crops. Even small children can contribute to the farmer's efforts, and the farmer's wife, relegated to household and kitchen garden chores, can care for several small children at one time without detracting greatly from her important contributions to the family's economy. Children, and the fertility of women in such societies, are highly valued; a new, but more precarious, equilibrium point is established between population numbers and environmental resources in many agricultural societies. Migration of peoples to distant lands and the extension of agricultural activities to undeveloped, even marginal, lands provides an outlet for population growth, one which had not characterized the population history of earlier groups who had been so immediately dependent on the existing biota found *in situ* by hunters and gatherers.

Over the past few hundred years, the industrial revolution and the scientific-technological changes which it engendered have appeared and spread unevenly and incompletely throughout the world. In those parts of the world where it first appeared, a frequently observed pattern of demographic change was associated with this revolution, characterized by the attainment first of low mortality and later low fertility rates. Some demographers optimistically predicted a demographic transition, extrapolated from this Euroamerican experience, would follow the industrialization and modernization of developing countries in the rest of the world. As Peterson (1975) has noted, however, the leaders of developing countries have often been motivated by a feeling that population growth is associated with power and prestige, and in some of these countries, declining mortality rates have not been followed by declining fertility rates. Migration no longer provides the viable outlet which it once did for some of the overpopulated nations in an interdependent modern world. It is, at best, unlikely that a new density-equilibrium system of human population numbers and environmental resources has yet been established in today's world.

The value and limitations of applying a demographic approach to the study of human adaptability and adaptation can now be examined in the light of this reconstruction of human population history. In order to elucidate the vital rates of anthropological populations, age-specific birth and death rates must usually be inferred from a census, since adequate direct data are rarely available and field workers do not usually remain in the area long enough to collect reliable long-term information. Some workers (Howell, 1976) have argued that it is possible to extrapolate from known demographic patterns of the present to past populations, on the principle that basic biological processes (ovulation, spermatogenesis, maturation, senility, etc.) continue to respond to environmental variations, and that for the last 40,000 years, at least, there has been no evolution of demographically relevant biological processes. Since a census is the result of birth and death rates which have operated in the past to produce the data collected in a census, then a life table, based on the assumption that these rates are fixed, can be constructed from census materials.

The life table, which has proven a powerful tool in a demographic analysis, is based on stable population theory. Basically, it is assumed that the age-specific vital rates used to develop a life table are the same for every cohort born in the population. Thus, if 10 of every 1000 children aged 0-1 years in 1960 will die in the following year, we assume that the same rate will apply to children belonging to the same cohort, aged 0-1 year, in 1961. This assumption does not, of course, require the additional assumption that we are dealing with a *stationary population*, one in which the population is neither increasing nor decreasing in size, but only that the age-specific vital rates for each cohort at a certain period will apply to the same aged group at a later period.

In the construction of a life table from complete data, a synthetic cohort, or *radix*, of 100,000 hypothetical individuals is exposed to the "risk" of dying, calculated from the age-specific mortality rates prevailing at that time for each age grouping. These tables are usually calculated for each sex separately, reflecting sex differences in mortality experiences, and are very frequently shown by five-year age cohorts (0-4, 5-9, 10-14, etc.). Thus it can be calculated, for example, how many of the members of the initial hypothetical cohort will be alive when the cohort has reached the ages of 40-44, 65-69, etc. A life table consists of a series of columns expressing for each cohort such attributes as the number of individuals living at exact age X, the probability of dying at age X, the number of individuals alive between age X and $X + 1$, and the expectation of life remaining to individuals who attain age X (see Table 11-1).

In fact, many anthropological populations have undergone major demographic disturbances, and Weiss (1975) has considered the probable effect of a range of disturbing events on the age distribution of hypothetical populations. From this study, he concludes that the most disruptive single type of disturbing event is a sudden jump in the number

TABLE 11-1 Life Table for Females in Chile, 1940

Age	Probability of Dying Between Age x and Age x + 1	Probability of Surviving Between Age x and Age x + 1	Number of Deaths Between Age x and Age x + 1	Survivors at Exact Age x	Years Lived Between Age x and Age x + 1	Total Years Lived After Exact Age x	Expectation of Life. Average Number of Years After Exact Age x
(1)	(2)	(3)	(4)	(5)	(6)	(7)	(8)
x	$1000\,q_x$	p_x	d_x	l_x	L_x	T_x	o_x
0	188.48	.81152	18,848	100,000	86,505	4,306,463	43.06
1	60.50	.93950	4,910	81,152	78,451	4,219,958	52.00
2	25.75	.97425	1,963	76,242	75,172	4,141,507	54.32
3	12.35	.98765	918	74,279	73,791	4,066,335	54.74
4	7.47	.99253	548	73,361	73,087	3,992,544	54.42
5	5.05	.99495	368	72,813	72,629	3,919,457	53.83
6	4.08	.99592	296	72,445	72,297	3,846,828	53.10
7	3.53	.99647	255	72,149	72,022	3,774,531	52.32
8	2.37	.99763	171	71,894	71,809	3,702,509	51.50
9	1.93	.99807	139	71,723	71,654	3,630,700	50.61
10	2.04	.99796	146	71,584	71,511	3,559,046	49.72
11	2.52	.99748	180	71,438	71,348	3,487,531	48.82
12	3.20	.99680	228	71,258	71,144	3,416,187	47.94
13	3.93	.99607	279	71,030	70,891	3,345,043	47.09
14	4.68	.99532	331	70,751	70,586	3,274,152	46.28
15	5.41	.99459	381	70,420	70,230	3,203,566	45.49
16	6.12	.99388	428	70,039	69,825	3,133,336	44.74
17	6.77	.99423	471	69,611	69,376	3,063,511	44.01
18	7.36	.99264	509	69,140	68,886	2,994,135	43.31
19	7.88	.99212	541	68,631	68,361	2,925,249	42.62
20	8.32	.99168	566	68,090	67,807	2,856,888	41.95
21	8.68	.99132	586	67,524	67,231	2,789,081	41.31
22	8.94	.99106	594	66,938	66,639	2,721,850	40.65
23	9.12	.99088	605	66,339	66,037	2,655,211	40.02
24	9.23	.99077	607	65,734	65,431	2,589,174	39.39
25	9.28	.99072	605	65,127	64,825	2,523,743	38.75
26	9.31	.99069	601	64,522	64,222	2,458,918	38.11
27	9.32	.99068	596	63,921	63,623	2,394,696	37.46
28	9.35	.99065	592	63,325	63,029	2,331,073	36.81
29	9.39	.99061	589	62,733	62,439	2,268,044	36.15
30	9.45	.99055	587	62,144	61,851	2,205,605	35.49
31	9.52	.99048	586	61,557	61,264	2,143,754	34.83
32	9.61	.99039	586	60,971	60,678	2,082,490	34.16
33	9.72	.99028	587	60,385	60,092	2,021,812	33.48

TABLE 11-1 (continued)

Age	Probability of Dying Between Age x and Age x + 1	Probability of Surviving Between Age x and Age x + 1	Number of Deaths Between Age x and Age x + 1	Survivors at Exact Age x	Years Lived Between Age x and Age x + 1	Total Years Lived After Exact Age x	Expectation of Life. Average Number of Years After Exact Age x
(1)	(2)	(3)	(4)	(5)	(6)	(7)	(8)
x	$1000\,q_x$	p_x	d_x	l_x	L_x	T_x	o_x
34	9.84	.99016	588	59,798	59,504	1,961,720	32.81
35	9.98	.99002	591	59,210	58,915	1,902,216	32.13
36	10.14	.98986	594	58,619	58,322	1,843,301	31.77
37	10.32	.98968	599	58,025	57,726	1,784,979	30.76
38	10.52	.98948	604	57,426	57,124	1,727,253	30.08
39	10.73	.98927	610	56,822	56,517	1,670,129	29.39
40	10.94	.98906	615	56,212	55,905	1,613,612	28.71
41	11.16	.98884	621	55,597	55,287	1,557,707	28.02
42	11.38	.98862	625	54,976	54,664	1,502,420	27.33
43	11.58	.98842	629	54,351	54,037	1,447,756	26.64
44	11.79	.98821	634	53,722	53,405	1,393,719	25.94
45	12.03	.98797	639	53,088	52,769	1,340,314	25.25
46	12.32	.98768	646	52,449	52,126	1,287,545	24.55
47	12.68	.98732	657	51,803	51,475	1,235,419	23.85
48	13.12	.98688	671	51,146	50,811	1,183,944	23.15
49	13.67	.98633	690	50,475	50,130	1,133,133	22.45
50	14.32	.98568	713	49,785	49,429	1,083,003	21.75
51	15.10	.98490	741	49,072	48,702	1,033,574	21.06
52	16.02	.98398	774	48,331	47,944	984,872	20.38
53	17.07	.98293	812	47,557	57,151	936,928	19.70
54	18.23	.98177	852	46,745	46,319	889,777	19.03
55	19.46	.98054	893	45,893	45,447	843,458	18.38
56	20.71	.97929	932	45,000	44,534	798,011	17.73
57	21.95	.97805	967	44,068	43,585	753,477	17.10
58	23.16	.97684	998	43,101	42,602	709,892	16.47
59	24.43	.97557	1,028	42,103	41,589	667,290	15.85
60	25.86	.97414	1,062	41,075	40,544	625,701	15.23
61	27.57	.97243	1,103	40,013	39,462	585,157	14.62
62	29.68	.97032	1,155	38,910	38,333	545,695	14.02
63	32.25	.96775	1,218	37,755	37,146	507,362	13.44
64	35.20	.96480	1,286	36,537	35,894	470,216	12.87
65	38.40	.96160	1,353	35,251	34,575	434,322	12.32
66	41.71	.95829	1,414	33,898	33,191	399,747	11.79
67	45.01	.95499	1,462	32,484	31,753	366,556	11.28
68	48.23	.95177	1,496	31,022	30,274	334,803	10.79
69	51.54	.90023	1,522	29,526	28,765	304,529	10.31
70	55.17	.94483	1,545	28,004	27,232	275,764	9.85
71	59.35	.94065	1,570	26,459	25,674	248,532	9.39

TABLE 11-1 (continued)

Age	Probability of Dying Between Age x and Age x + 1	Probability of Surviving Between Age x and Age x + 1	Number of Deaths Between Age x and Age x + 1	Survivors at Exact Age x	Years Lived Between Age x and Age x + 1	Total Years Lived After Exact Age x	Expectation of Life. Average Number of Years After Exact Age x
(1)	(2)	(3)	(4)	(5)	(6)	(7)	(8)
x	$1000\, q_x$	p_x	d_x	l_x	L_x	T_x	$\mathring{e}_x$
72	64.30	.93570	1,600	24,889	24,089	222,858	8.95
73	70.14	.92986	1,634	23,289	22,472	198,769	8.53
74	76.60	.92341	1,659	21,655	20,826	176,297	8.14
75	83.24	.91676	1,665	19,996	19,164	155,471	7.78
76	89.70	.91030	1,644	18,331	17,509	136,307	7.44
77	95.56	.90444	1,595	16,687	15,890	118,798	7.12
78	100.60	.89940	1,518	15,092	14,333	102,908	6.82
79	105.27	.89473	1,429	13,574	12,860	88,575	6.53
80	110.18	.88982	1,338	12,145	11,476	75,715	6.23
81	115.96	.88404	1,253	10,807	10,181	64,239	5.94
82	123.23	.87677	1,177	9,554	8,966	54,058	5.66
83	132.37	.86763	1,109	8,377	7,823	45,092	5.38
84	142.87	.85713	1,038	7,268	6,749	37,269	5.13
85	153.97	.84603	959	6,230	5,751	30,520	4.90
86	164.91	.83509	869	5,271	4,837	24,769	4.70
87	174.94	.82506	770	4,402	4,017	19,932	4.53
88	183.47	.81653	666	3,632	3,299	15,915	4.38
89	190.50	.80950	565	2,966	2,684	12,616	4.25
90	196.17	.80383	471	2,401	2,166	9,932	4.14
91	200.66	.79933	387	1,930	1,737	7,766	4.02
92	204.12	.79588	315	1,543	1,386	6,029	3.91
93	206.85	.79315	254	1,228	1,101	4,643	3.78
94	209.67	.79033	204	974	872	3,542	3.64
95	213.59	.78642	164	770	688	2,670	3.47
96	219.56	.78044	133	606	540	1,982	3.27
97	228.57	.77143	108	473	419	1,442	3.05
98	241.88	.75812	88	365	321	1,023	2.80
99	261.84	.73816	73	277	241	702	2.53
100	291.08	.70892	59	204	175	461	2.25
101	332.25	.66775	48	145	121	286	1.97
102	349.63	.65037	38	97	78	165	1.70
103	401.87	.59813	27	59	46	87	1.47
104	454.86	.54514	17	32	24	41	1.28
105	496.49	.50351	9	15	11	17	1.13
106	514.72	.48528	4	6	4	6	1.00
107	–	–	–	–	–	–	–

From Barclay, 1958.

of children born, for the effect of an increase in cohort at the base of a population pyramid is expressed at every subsequent stage in the pyramid so long as they live.

A modern example of this phenomenon has been described by Clark (1974), based on Teitelbaum's studies of the fertility effects of the abolition of legal abortion in Romania, published in *Population Studies* (1972). The Romanian government, through Government Decree No. 463 issued in 1957, had made abortion readily and inexpensively available to all women whose pregnancies had not passed the twelfth week of gestation. Abortion had become the primary means of family size limitation, and the fertility rate had steadily declined over the years. But, as indicated in Figure 11-1, a radical change in fertility patterns occurred within eight months of the issuance of Decree No. 770–banning all abortions except in the most extreme cases. Within eleven months after the issuance of

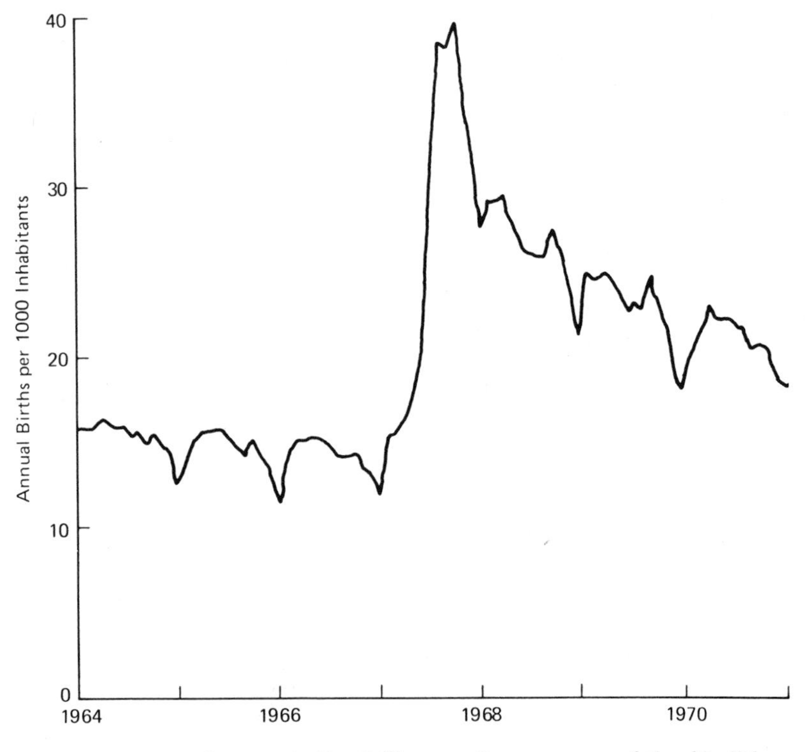

FIGURE 11-1 Changes in Fertibility as a Consequence of the Abolition of Legal Abortion in Romania (adapted from Clark, 1974.)

this latter decree, the birth rate had tripled; and although it began to decline shortly after this time, the cohorts born in 1967 and 1968 were double the size of the cohorts which had preceded them. The sudden abolition of the government's pro-abortion policy produced a cohort of radically increased numbers born in 1967 and 1968. As Clark points out, the school system must adjust to handle this sudden and short-lived influx of children. Given prevailing marital-age patterns, females belonging to this age group as adults must find husbands from the much smaller cohorts born earlier, and males, on reaching adulthood, must seek wives from the smaller cohorts born after 1968. If the greater numbers of those born in 1967 and 1968 continue the fertility patterns of older cohorts, a secondary bulge in numbers should appear in cohorts yet to be born. How will employers find positions for the large number of future workers born in 1967 and 1968? And, as this group approaches retirement age in the future, how much additional cost will their support place on the smaller cohorts which follow them?

In contrast, Weiss found that other single events (epidemic, infant disease, reduction in number of producing adults with total population increase, etc.) of limited effective duration have minimal long-term effects on numbers or age distribution. However, disruptive events of five-year effective duration which occur every decade, or a series of disruptive events over a ten-year effective period, would so seriously distort the population as to make the use of stable population models untenable. While perturbations of this magnitude are relatively uncommon they are not unknown. To take just one example, McArthur reports the following sequence of events over one thirty-year period in the history of Tonga: an epidemic which led to the deaths of "hundreds" in 1797; an outbreak of civil war in 1799; and a severe famine during the entire period of 1799–1826. Commendably, Weiss notes that populations which show evidence of extensive demographic disturbance, as revealed either in anecdotal evidence or in skewed age distributions, are not suitable for life-table analysis without suitable adjustment. Nonetheless, he argues that by the use of underlying average rates, applied to populations which have not been severely disrupted demographically, a model life table can be developed to provide a productive approach to the study of human adaptation.

Just as the use of the Hardy-Weinberg model allows us to compare observed and predicted results and to test for the causes of discrepancies between the real world and simulated conditions, we can perhaps best use the model life table to detect and test for the causes underlying perturbations in the composition and structure of real populations. For example, Neel and Weiss (1975) calculated a life expectancy of twenty years from the actual mortality experience of the Yanomamo of South America. When compared to model life tables, a better fit of population data is found with a life-table model for life expectancy of thirty years at birth, except that mortality seems to be higher among infants and young adults than predicted by the model. In part, this discrepancy is probably related to the Yanomamo practice of infanticide; further investigations will have to pursue the specific causes of mortality among young adults and to reconstruct the demographic history of the young adults studied by Neel and his colleagues.

Any human population which exploits to the point of depletion the resources available to it risks extinction. Thus, any mechanism which reduces this risk enhances the probability of survival. If population size is maintained at an equilibrium level, this danger can be minimized; and so far as we can now infer, most ancient populations of hunters and gatherers maintained their numbers below the "carrying capacity" of the environment. This seems to have been effected through a combination of moderate fertility and high mortality rates for the greater part of human history. Several physiological explanations have been advanced to account for the moderate fertility levels reported from contemporary groups of hunters and gatherers and, by extension, for prehistoric populations. These include the inhibiting effects of lactation on ovulation and the possible inhibition of ovulation which is claimed to result from a lack of minimal fat deposits in females. Both of these processes are subject to and may be supplemented by cultural influences.

For one thing, many societies sanction extended periods of breastfeeding, and in a number of societies sexual intercourse is prohibited during the entire period when the child is being nursed. The inhibiting effects of lactation on ovulation appear to be most effective when the child is obtaining all or nearly all of its nourishment from mother's milk. When supplementary foods are introduced into the child's diet, and nursing becomes less intense, the mother seems to produce lower levels of the hormone prolactin which, with other hormones, acts to suppress ovulation. Then, too, in societies which allow the mother to reduce her daily activities or increase her caloric intake beyond the extra metabolic demands of nursing, new mothers may more readily maintain or resume the minimal body fat composition which must be established before ovulation can recommence.

In any case, a number of behavioral patterns are directly relevant to maintaining moderate fertility levels in simple societies. Prime among these is infanticide, which operates after a woman has been removed from the risk of pregnancy for at least the better part of a year. The only more effective control on fertility at this technological level, short of complete abstinence, is a high mortality rate among young children. In the latter case, the mother of a child who dies at or near the end of the period during which it has been suckled has been at minimal risk of pregnancy for several years; so, on the average, fertility would have been minimized among women in populations characterized by lengthy periods of breast feeding and high rates of mortality in weanlings.

Critical to the adaptive value of these conditions is the ability of populations to adjust quickly to changes. Since several of the major limitations on fertility patterns of economically simple societies involve biophysiological and cultural factors, responses to environmental changes can be quickly and readily made. The loss of a child who is still nursing removes a physiological restraint (prolactin production) on ovulation, and removes any cultural barrier (intercourse tabus) on the resumption of sexual activity. A number of societies which place sexual tabus on relatives of the dead often do not extend those restrictions to the parents of young children who have died. Access to new or greater amounts of resources—because of technological advances which enhance the availability of local resources, because of migration of the human group to a better setting, or as a result of improvements in the local biota, or even a better system of food distribution to females—can increase the fecundity of females through nutritional improvements, and allow the population to attain a higher fertility level compatible with the new economic situation.

Various kinds of feedback relationships between available resources and fertility mechanisms can provide human populations with means of responding to environmental stress. The frequency of infanticide can be increased, post-partum tabus more scrupulously followed, and the nursing period extended when resources become scarce. Populations can become more dispersed and mobility increased in a wider search for scarce foods, requiring additional caloric expenditures for a lower caloric return, and placing additional stress on female body fat deposits. The absence of hunters for longer periods can reduce the opportunity for sexual activity, and nonproductive elderly individuals may be sacrificed to the needs of the active adults and children.

Some of these responses are not so readily available to more sedentary populations, and ideological or religious sanctions supportive of high fertility found in some of these populations are notoriously difficult to reverse. In many of these preindustrial societies, high fertility levels are matched by high mortality levels; in a kind of vicious circle, the experience of high rates of mortality among children seems to inhibit the adoption of family-planning practices in contemporary traditional countries, even where the prevailing ideology is ambivalent toward the adoption of modern contraceptive practices. In an indirect way, another trend may also act to inhibit the acceptance of contraception—when women in developing countries adopt bottle-feeding of their children, thus removing one constraint (even if it is not a 100 percent effective method) on fertility. As Knodel (1977) points out, bottle feeding in many of these societies, as in certain European populations in the late nineteenth and early twentieth centuries, is associated with an increase in infant mortality; so, the effects on population size are not immediately so great as could be expected. But the repeated experience of having children die can restrain individuals and families from adopting contraceptive practices, lest they risk leaving behind no children at all to continue the family line, inherit property, to carry out necessary rituals for the continued well-being of their dead ancestors, or to care for aging parents.

But sedentary life, the increased population density of agricultural communities, entails new kinds or more severe intensities of disease selection. The high rates of mortality from epidemic diseases in settled populations must have had a critical impact on genetic variation: relatively few would have survived to reach reproductive maturity, although fertility was high. Where selection operates most restrictively in the prereproductive years, genetic variability is reduced, and a wide range of causes of selection has little opportunity to be expressed. The effects on population size and age distributions of populations are probably minimal if Weiss's studies on the effects of disturbing demographic effects can be applied here. But this kind of "hard selection" would have a strong effect on genetic variation. In addition, any tendency towards endogamy within settled communities and regions may limit gene flow and increase levels of inbreeding above those reported for a number of contemporary populations of hunters and gatherers. Finally, as social and economic differences grow out of the accumulation of land, property, and goods, social stratification may have been accompanied by class and caste distinctions. The accompanying marital restrictions would further segment populations into co-resident reproductive isolates with distinctive gene pools. For most of this preindustrial period of human history, the number of the species was maintained at higher levels than ever before; yet human populations evolved, to large degree, in relative isolation from one another. Only during the latter part of this period did contact begin between explorers and the native peoples throughout the world to whom these intrepid visitors introduced their diseases and their genes.

Demographers tell us that the population explosion associated with the industrial revolution and its aftermath entailed, first of all, a reduction of mortality—which followed the application of simple rules of hygiene and sanitation. John Graunt, usually acknowledged as the father of demography, analyzed the weekly reports of parish clerks on burials and christenings in seventeenth-century London to discover that deaths exceeded births in that great city; and this was probably true of most cities of that time. Cities of this period depended on migration from rural areas for their continued growth. In England and Wales, the birth rate began to exceed the death rate only in the nineteenth century; McKeown and Record (1963) argue that the decline of mortality there in the latter half of the nineteenth century was due to improved diet, hygienic changes introduced by sanitary reformers, and coadaptation of certain pathogens (for scarlet fever, tuberculosis, typhus, and cholera) and their human hosts. Since the advent of vaccination, innoculations against various diseases, and the availability of antibiotics, mortality rates have been radically decreased, at first in Western Europe and the United States, and most recently among populations throughout most of the rest of the world. In many areas, a far larger proportion of all those born survive to reach the age of reproductive maturity, and a kind of "soft selection" can operate in which a variety of causes selects among adults who die at various points in their reproductive lifespans.

A greater genetic variability exists and can be expressed in offspring who are, themselves, subject to a wide variety of selective agencies. At the same time, greater conscious control of fertility is available than has ever been the case in the past. As we shall see, the mortality component of the index for the opportunity of selection is becoming less significant at the same time that the fertility component assumes greater importance. In industrialized countries, even slight variations in fertility assume critical selective value.

But what of demographic responses to these changes? The doomsayers would have us adopt lifeboat ethics in a world teetering on the edge of ecological disaster. It seems clear that we can never return to the equilibrium levels that our early ancestors seem to have maintained; and who among us is willing or able to reinstitute some of those practices—infanticide, cannibalism, senilicide—which once helped to assure that the numbers of people would not exceed the ability to survive on the resources which could be hunted or gathered? Shall we institute a policy of starving young girls and new mothers, in the hope that this will reduce their fecundity? Just as some ancient or historical populations must have exceeded their resources temporarily, the population of today's world is undergoing the stress of disequilibrium. Tomorrow's discoveries may change the amounts and kinds of resources available to us. Solar energy or nuclear energy plants may alleviate the growing shortage of fossil fuels, just as coal once replaced dwindling supplies of timber resources for heating and other energy requirements. The "Green Revolution," which has disappointed some authorities, is not yet over; genetic research may yet lead to the alleviation of food shortages. And human behavior is changeable: members of affluent societies are reducing their consumption of fats and sugars; our thermostats are being set lower, and increasing numbers are purchasing small, low gasoline-consuming cars.

Perhaps more important, humans are consciously able to control their fertility more reliably than was ever possible in the past. If we can learn more of the motivations and conditions which make people willing and able to control their fertility, we can effect the necessary changes in our reproductive behavior to match corresponding changes in mortality patterns. If we can raise the standard of living for all peoples, so that they can expect to see most of the children they bear survive them, and reduce tensions between nations whose people share fairly in the world's economic resources, a great deal of the motivation for pronatalist attitudes would be removed. Hopefully, we have come too far to exclude other humans from a lifeboat, or to be forced on a lifeboat unnecessarily when our capable ship is only sailing briefly through a rough patch of seas.

REFERENCES AND RECOMMENDED READING

ACSÁDI, G., and J. NEMESKERI 1970. *History of Human Lifespan and Mortality.* K. Balas, translator. Budapest: Adadémiai Kiadó.

BARCLAY, G.W. 1958. *Techniques of Population Analysis.* New York: John Wiley and Sons, Inc.

BILLEWICZ, W.Z., H.M. FELLOWES and C.A. HYTTEN. 1976. Comments on the critical metabolic mass and the age of menarche. *Annals of Human Biology,* 3(1):51–59.

BIRDSELL, J.B. 1968. Some predictions for the Pleistocene based on equilibrium systems among recent hunters-gatherers. In *Man the Hunter.* R.B. Lee and I. DeVore, eds. Chicago: Aldine Publishing Company. Pp. 229–240.

——. 1957. Some population problems involving Pleistocene man. *Cold Spring Harbor Symposia in Quantitative Biology* 22:47–69.

BOGUE, D.J. 1969. *Principles of Demography.* New York: John Wiley and Sons, Inc.

CLARK, L. 1974. Baby boom by fiat: the effects of population policy on fertility in Romania. An Inquiry Teaching Module, *Teaching Notes on Population,* III:24–39.

DUNN, L. 1968. Epidemiological factors: health and disease in hunter-gatherers. In *Man the Hunter.* R.B. Lee and I. DeVore, eds. Chicago: Aldine Publishing Company. Pp 221–228.

JOHNSTON, F.E. 1973. *Microevolution of Human Populations.* Englewood Cliffs: Prentice-Hall.

JOHNSTON, F.E., R.M. MALINA and M.A. GALBRAITH. 1971. Height, weight and age at menarche and the "critical weight" hypothesis. *Science* 174:1148.

HOWELL, N. 1976. Toward a uniformitarian theory of human paleodemography. In *The Demographic Evolution of Human Populations.* R.H. Ward and K.M. Weiss, eds. New York: Academic Press. Pp. 25–40.

KNODEL, J. Breast-feeding and population growth. *Science* 198:1111–1115.

KRZYWICKI, L. 1934. *Primitive Society and its Vital Statistics.* London: MacMillan and Co., Ltd.

LEE, R.B. 1968. What hunters do for a living, or, how to make out on scarce resources. In *Man the Hunter.* R.B. Lee and I. DeVore, eds. Chicago: Aldine Publishing Company. Pp. 30–48.

McARTHUR, N. 1968. *Island Populations of the Pacific.* Honolulu: University of Hawaii Press.

McKEOWN, T., and R.G. RECORD 1963. Reasons for the decline of mortality in England and Wales during the nineteenth century. *Population Studies* 16:94–122.

NAG, M. 1962. Factors affecting human fertility in nonindustrial societies: a cross-cultural study. *Yale University Publications in Anthropology,* No. 66.

NEEL, J.V., and K.M. WEISS 1975. The genetic structure of a tribal population, the Yanomama Indians. *American Journal of Physical Anthropology* 42:25–52.

PETERSEN, W. 1975. *Population.* 3rd edition. New York: Macmillan Publishing Company.

UNDERWOOD, J.H. 1973. The demography of a myth: abortion in Yap. *Human Biology in Oceania* 2:115–127.

VAN'T HOF, M.A., and M.J. ROEDE. 1977. A Monte Carlo test of weight as a critical factor in menarche, compared with bone age and measures of height, width and sexual development. *Annals of Human Biology* 4(6):581–586.

WEISS, K.M. 1976. Demographic theory and anthropological inference. In *Annual Review of Anthropology,* 1976, Vol. 5. B.J. Siegel, A.R. Beals and S.A. Tyler, eds. Palo Alto: Annual Reviews, Inc. Pp. 351–382.

——. 1975. Demographic disturbance and the use of life tables in anthropology. In *Population Studies in Archaeology and Biological An-*

thropology, a Symposium. A.C. Swedlund, ed. *Memoirs of the Society for American Archaeology,* No. 30. Pp. 46–56.

YENGOYAN, A. 1968. Demographic and ecological influences on aboriginal Australian marriage sections. In *Man the Hunter.* R.B. Lee and I. DeVore, eds. Chicago: Aldine Publishing Company. Pp. 185–199.

CHAPTER 12

The Ongoing Microevolution of Human Populations

A few years ago a young graduate student from the Genetics Department requested an appointment to discuss with me her ongoing research on the topic, "artificial selection among human populations." When she arrived in my office I quickly explained that I was unfamiliar with the subject of artificial selection, except in the Darwinian sense of attempts by humans to control the breeding of plants and animals. It quickly became apparent that she was using the phrase in a very different way. Since natural selection in the animal populations with which she was familiar involved the action of selective agents over which the subject populations had no conscious control, she felt that a unique kind of selection was affecting human populations. The entire conversation epitomized a viewpoint which is not uncommonly expressed by those who envision humans as now removed, either by creation or through evolution, from the direct operation of ordinary evolutionary processes. The argument is sometimes made that culture is free of organic influence or causation, and that our cultural heritage has somehow superseded biological laws and processes.

Nothing could be further from the truth, of course, for culture cannot exist independent of biological organisms with the complex neurophysiological systems and anatomical structures required for the communication through language of those socially transmitted and shared ideas, knowledge, values, attitudes, patterns of behavior, and customs which constitute cultural phenomena. Just as Mendelian populations have certain attributes such as gene pools, which cannot be understood wholly in terms of the properties (genomes) of individuals, cultural systems also have properties such as transmissibility, which attach to human societies. This does not remove cultural systems from biological influences or processes, despite the fact that cultural content can be transmitted without becoming part of the genetic heritage of human populations.

In most cases, however, this anti-biological viewpoint proceeds more directly from observations of the great advances which have been made in the control of some aspects of the human environment. Smallpox epidemics, which once decimated human populations, may already be the province of medical historians. Poliomyelitis appears only sporadically among the relatively rare, uninoculated members of populations with adequate public health programs. Hepatitis outbreaks occur infrequently among the residents of cities and towns that maintain basic standards and systems for waste disposal and water purification. Our conscious efforts, the development and application of scientific knowledge and principles, have virtually eliminated certain agents of selection at least in some parts of the world. But, in all honesty, what conditions made these diseases effective agents of selection in the past? Weren't they dependent on the presence of densely populated settlements of human beings, which cultural developments made possible?

One example of "relaxed selection" in human populations, which is frequently cited in textbooks, refers to the incidence of color-blindness in populations living under varied ecological circumstances. Although several forms of color blindness are known, involving different genetic loci and producing different perceptive anomalies, the frequency of color blindness in general is lower in groups of hunters and gatherers than among sedentary agriculturalists, and is highest among populations of industrialized nations or urban centers. The usual explanation of this distribution pattern centers on the presumably greater selective value of color vision among peoples who would be at risk if they collect and eat a poisonous plant distinguishable only by its color, or if they hunt a potentially dangerous animal which cannot be distinguished from the surrounding foliage, its natural camouflage, if its colors are not perceived. This argument neatly fits the observed incidence of the genetic trait. However, little attention has been given to an empirical demonstration of its validity, and one would like to know if any of the plants which are collected by foragers are distinguishable only by color from dangerous varieties, or if hunters are really jeopardized by imparied color perception. Does it really matter what the color of a charging elephant is? Rather, as one who suffers from one form of color blindness, I have risked life and limb on more than one occasion when I attempted to connect what must have been red wires to green terminals in electrical appliances. And, while I have learned that the red light is on the top of most traffic light standards, I once nearly caused mayhem in a small Mexican village in which the traffic lights were reversed from the usual red-yellow-green positioning with which I am familiar.

Cruz-Coke (1970), on the other hand, has suggested that alleles at the loci involved in color blindness may have pleiotropic effects and that these pleiotropic effects may be the targets of selection. In Chile, Cruz-Coke and his co-workers claim to have found an association between alcoholism and a defect in blue-yellow vision. This association has not been confirmed in some other studies, but the hypothesis that selection against traits which are produced by the pleiotropic effects of alleles responsible for color vision defects merits further attention.

But if selection has lessened in respect to some traits, and there is good evidence that this is so, this is not to say that natural selection no longer operates on human gene pools. Indeed, the cultural heritage which has made the maintenance and control of some diseases possible has also been responsible: for the production of chemical agents, such as thalidomide, which adversely affect the normal development of the fetus; for chemical materials which seem to inhibit the production of gametes; for vaccines which appear to increase the incidence of certain diseases other than those against which they are intended to provide protection; and for carcinogenic agents, materials which increase the risk of developing cancer. Even the medical technology which is applied to save lives can increase mutation rates, and the military technology designed to win wars can alter the gene pools of friend and enemy alike. Natural selection, as well as the other evolutionary processes, continues to operate in contemporary human populations.

You can hardly read a newspaper, magazine, or journal today without finding at least one or more articles concerned with the biological consequences of changing lifestyles among the peoples of industrialized nations. Up to 30 percent of Americans in the United States are obese, weighing at least 30 percent more than they should, and obesity has been associated with such adverse effects on health as an increased incidence of strokes, heart disease, and hypertension (Kolata, 1977). We are said to be suffering from affluent malnutrition, our diets loaded with animal fats, refined foods, sugars, and starches. Until recently, babies were encouraged to eat even more by offering them prepared baby food flavored with monosodium glutamate, at least until animal studies indicated that the compound might damage the brain tissues of young animals. We treat other foods with preservatives such as nitrites, which show mutagenic effects—at least when applied to the raw fish that form an important part of the Japanese diet, providing a possible explanation of the higher frequency of stomach cancer among Japanese than among Americans, who are more likely to eat nitritie-treated beef and hot dogs (Marquardt et al., 1977). Responding to warnings of the dangers of obesity, we turn increasingly to sugar substitutes, saccharin and cyclamates, despite animal studies indicating their possible carcinogenic effects, or we turn to inadequate liquid protein diets which have now been directly linked to a number of deaths.

In an effort to maintain the youthful appearance which is so highly valued among many of these societies, men and women, perhaps 30 million in the United States alone, use hair dyes which may contain suspected carcinogenic substances. And, despite the Surgeon General's warnings, we continue to smoke tobacco—a practice which Retherford (1975) argues is responsible for most of the increase in summary measures of sex mortality differentials in the United States during this century, and which has made a substantial contribution to the presently high incidence of widowhood among older women. Recent reports indicate that tobacco smoking is increasing, particularly among teenage girls in the United States; and if this behavior is continued into their reproductively active years, we can expect to see an increase in low birth-weight babies, since the reported association between smoking and low birth weight has now been confirmed in a number of studies. These babies are, as discussed in Chapters 7 and 9, less likely to survive the critical neonatal period of life.

Nowhere are the biologically relevant changes in life patterns more clearly evident than in certain demographic features. In the industrialized countries, particularly of North America and Western Europe, far-reaching changes in fertility and mortality patterns have taken place over the past century. As Richardson and Stubbs (1976) point out, in 1840 only about half the women ever born produced children, whereas today about 85 percent of all women reproduce. In 1880, less than 75 percent of all individuals reached the age of 15 years and less than 70 percent attained the age of 30 years. In contrast, over 95 percent of all individuals attained the age of 30 years in 1960. A far greater percentage than was the case only one hundred years ago now survive to reproduce, and this must enhance the variation present in the gene pools of these populations.

This change has involved radical shifts in mortality patterns, and particularly but not exclusively in patterns of infant and childhood mortality. A comparison of the ten leading causes of death in the United States in 1900 and 1967 (Table 12-1) reflects the decreasing importance of certain diseases as major agents of natural selection. Diphtheria and meningitis, once responsible for over 4 percent of all deaths in this country, are no

TABLE 12-1 Ten Leading Causes of Death in the United States: 1900 and 1967

1900	*Deaths per 100,000 Persons*	*% of All Deaths*
1. Pneumonia and influenza	202	11.8
2. Tuberculosis	194	11.3
3. Diarrhea and enteritis	143	8.3
4. Diseases of the heart	137	8.0
5. Cerebral hemorrhage	107	6.2
6. Nephritis	89	5.2
7. Accidents	72	4.2
8. Cancer	64	3.7
9. Diphtheria	40	2.3
10. Meningitis	34	2.0
1967		
1. Diseases of the heart	365	39.0
2. Cancer	157	16.8
3. Cerebral hemorrhage	102	10.9
4. Accidents	57	6.1
5. Pneumonia and influenza	29	3.1
6. Certain diseases of early infancy	24	2.6
7. General arteriosclerosis	19	2.0
8. Diabetes mellitus	18	1.9
9. Other diseases of the circulatory system	15	1.6
10. Emphysema and related diseases	15	1.6

Wallace, B. 1972. *Essays in Social Biology,* vol. III: *Disease, Sex, Communication, Behavior.* Englewood Cliffs, N.J.: Prentice-Hall, Inc. By permission of the publisher.

longer major health problems. Instead, heart diseases have advanced from fourth to first position and cancer from eighth to second, together accounting for well over 50 percent of all deaths in the United States in 1967, although they were responsible for less than 12 percent of deaths in 1900. In part, this reflects the changing age structure of the census population of the United States, since far more of us survive long enough to succumb to these conditions instead of dying as children from diphtheria. But changes in lifestyle, including the pursuit of sedentary, stressful occupations and improper diet in the case of heart diseases, and smoking and exposure to environmental pollution in the case of cancer, seem to be implicated in these trends.

While these changes have been taking place, other kinds of developments affecting the genetic heritage of human populations in the more economically affluent societies have also been taking place. Some of the partial barriers to intermixture have become less effective in the United States since the end of World War II, as changing attitudes have accompanied the fall of legal barriers supporting discrimination in employment, marriage, educational opportunities, and housing. Restrictive immigration policies, beginning at an earlier period in this century, have also contributed to a decrease in cultural and biological heterogeneity. Opposing this trend, however, is a very recent phenomenon, the growth of ethnic separatism. Members of various minority groups, while continuing to seek legal and quasi-legal redress against restrictions on participation in many aspects of the larger society, have begun to emphasize certain of the traditional patterns and customs of the native cultural heritage, to develop pride in native ancestry, and to revitalize ancestral lifestyles. It is yet too early to say if this trend will lead to a decrease in gene flow between ethnic or racial groups; but if this should happen, we can expect to see a retention, perhaps even an intensification, of genetic heterogeneity in multi-ethnic societies.

Another kind of change which has affected the industrialized nations involves the breakdown of regional and social isolation. With industrialization, improved means of transportation became available to even the more remote parts of nations, while economic opportunities for wage employment in urban and industrial centers resulted in a strong migratory "pull" effect. The consequences of these changes were to end regional isolation, to extend the geographical range from which potential mates could be sought, and to bring young people from different parts of the world together in metropolitan complexes. In time, social mobility, abetted by the opening of educational opportunities to a wider range of society, the need for new kinds of professionals, and the opportunities for material success available to individual entrepeneurs, resulted in an increased rate of gene flow between members of different social classes. Beset by the problems of a limited number of eligible candidates of the proper age and sex, by the presence of the allele for hemophilia among some branches of royal families, and by the declining fortunes of other lines, even European royalty in countries which maintained monarchies began to accept commoners as spouses and to reach into secondary lines of the remaining royal families for mates. Whatever genetic differences may have once characterized these kindreds have now begun to fade.

Many observed biological changes in the populations of industrialized countries owe more to the effects of changing attitudes than to altered mating patterns. The dietary changes which seem to underlie the secular trend toward increased stature and the declining age at menarche in populations of industrialized countries might have remained confined to economically advantaged groups in these societies if other social classes had not obtained improvements in economic status. Social legislation ended the wage employment of young children from the lower socioeconomic classes in deplorable circumstances in factories and industry. Labor and trade unions won hard battles for a greater share in the nation's wealth and for the development of programs for unemployment compensation, seeking to guarantee that minimal standards of living will be available to workers and their families, even under adverse conditions. In some, but not all, industrialized nations, adequate medical care is being made available to all through national health programs or subsidized medical insurance plans. Through such means, the cost of surgery, lengthy hospitalization, or the care of the terminally ill does not impose a catastrophic burden on individuals or their families, and every citizen is considered to possess the right to adequate medical care.

In the United States, as in most other industrialized nations, government has responded to public demands to regulate conditions which can adversely affect the health of all citizens. Since the early years of this century, minimal safety standards and working conditions have come under an increasing degree of regulation, food-processing plants are required to maintain minimal sanitary and health standards, drugs are now required to undergo prolonged testing before being released for public use. Without doubt, such measures have contributed to improved health, to a decrease in the rates of occupational diseases and disabilities, and to a decline in the incidence of poisoning and death from the consumption of contaminated foods.

In most of the industrialized countries, attitudes toward human fertility have also changed, with dramatic consequences. In England and in the United States, early advocates of birth control were subject to imprisonment and fines. Few of these pioneers lived long enough to see their rather moderate proposals win general approval and be surpassed by the acceptance of even more comprehensive programs for family planning. Only in recent years has elective abortion been described as a human right in some of these societies. The adoption of modern reliable contraceptive methods has not spread uniformly through the populations of these societies, although governmental and private agencies have accelerated the spread of family-planning information and contraceptive technology to members of economically disadvantaged groups who would otherwise have limited access to private medical facilities. The results have been sharp reductions in fertility rates, especially since the postwar "baby boom" years. In the United States, it is not certain if this represents a definitive trend or merely a temporary pattern of behavior among young women who are postponing reproductive performance to a later period of their lives (Sklar and Berkov, 1975). In the latter case, we might expect an increase in the frequency of congenital abnormalities, such as Down's syndrome, which show a maternal-age effect.

In contrast, teenage pregnancies are increasing, a fair percentage of them involving unmarried females, and this is clearly the age group which is not receiving the benefit of family-planning counseling before beginning sexual activity. A decrease in the average age at first pregnancy is also apparent among this group, and several factors are probably involved. It may be true that changing attitudes toward premarital intercourse affect this trend, but the decreasing age at menarche and a shorter period of adolescent sterility must also contribute significantly to this phenomenon. Since there is evidence that very young mothers, and their babies, are a high-risk category, changes in maternal mortality rates can be expected, as well as increases in the stillbirth and neonatal mortality rates.

The greater availability of elective abortions to women in countries or states which sanction medical abortion could also be expected to have significant biological consequences. Already, a rather impressive and constantly growing number of biochemical disorders can now be diagnosed prenatally (Table 12–2). Unfortunately, the diagnostic technique for many of these defects, amniocentesis, requires taking a sample of amniotic fluid and carries a slight risk to the fetus, so the procedure is not routinely practiced. For genetic defects, the appearance of the trait in relatives or in a previous child is usually required before the procedure will be employed. For conditions such as Tay-Sachs disease where heterozygous parents can be identified, prenatal diagnosis is also indicated. For parents of unborn children diagnosed as having one of these defects, elective abortion, particularly in the case of any one of the more debilitating conditions, can alleviate the severe economic, emotional, and social costs of rearing an affected child.

At first it would seem that the genetic effect of such practices on sublethal traits would be to reduce the frequency of the allele in the gene pool of the next generation. However, in the case of deleterious traits expressed by a homozygous recessive genotype, allele frequencies may not change in the expected direction if the parents replace the lost child, or "reproductively overcompensate," by giving birth to a greater number of unaffected heterozygous children to replace the lost child than they would otherwise have produced. Hartung and Ellison (1977) have even argued that if medical attention can increase the life expectancy of children with genetic disabilities due to a homozygous recessive genotype, without significantly increasing their reproductive capacities, parents are less likely to compensate by producing heterozygous children, thus lowering the equilibrium frequency of the allele in the gene pool!

TABLE 12-2 Biochemical Disorders That Have Been Diagnosed Prenatally

Disorder	*Metabolic Defect*
Acid phosphatase deficiency	Lysosomal acid phosphatase deficiency
Adenosine deaminase deficiency (combined immunodeficiency)	Adenosine deaminase deficiency
Adrenogenital syndrome	C-11 or C-21 steroid hydroxylase deficiency
Argininosuccinic aciduria	Argininosuccinase deficiency
Citrullinemia	Argininosuccinic acid synthetase deficiency
Cystinosis	Cystine accumulation
Fabry disease	Ceramidetrihexoside α-galactosidase deficiency
Fucosidosis	α-Fucosidase deficiency
Galactokinase deficiency	Galactokinase deficiency
Galactosemia	Galactose-1-phosphate uridyltransferase deficiency
Generalized gangliosidosis (G_{M1} gangliosidosis, Type I)	β-Galactosidase deficiency
Juvenile gangliosidosis (G_{M1} gangliosidosis, Type II)	β-Galactosidase deficiency
Gaucher disease	Glucocerebrosidase deficiency
Glycogen storage disease, Type II (Pompe disease)	α-1, 4-Glucosidase deficiency
Hemoglobinopathy (sickle cell)	Synthesis of hemoglobin S
Hunter syndrome	α-L-Iduronic acid-2 sulfatase deficiency
Hurler syndrome	α-L-Iduronidase deficiency
Hypophosphatasia (some types)	Alkaline phosphatase deficiency
I-cell disease	Multiple lysosomal enzyme deficiencies
Isovaleric acidemia	Isovaleryl CoA dehydrogenase deficiency
Ketotic hyperglycinemia	Propionyl CoA carboxylase deficiency
Krabbe disease	Galactocerebroside β-galactosidase deficiency
Lesch-Nyhan syndrome	Hypoxanthine guanine phosphoribosyltransferase deficiency
Maple syrup urine disease	Branched chain ketoacid decarboxylase deficiency
Maroteaux-Lamy syndrome	Arylsulfatase B deficiency
Menkes disease	Copper accumulation
Metachromatic leukodystrophy	Arylsulfatase A deficiency
Methylmalonic aciduria	Methylmalonic CoA mutase deficiency
Niemann-Pick disease	Sphingomyelinase deficiency

TABLE 12-2 (continued)

Disorder	*Metabolic Defect*
Placental sulfatase deficiency	Placental sulfatase deficiency
Porphyria–acute intermittent type	Uroporphyrinogen 1 synthetase deficiency
Pyruvate decarboxylase deficiency	Pyruvate decarboxylase deficiency
Sandhoff disease	Hexosaminidase A and B deficiency
Sanfilippo syndrome, Type A	Heparin sulfatase deficiency
Sanfilippo syndrome, Type B	N-acetyl-α-glucosaminidase deficiency
Tay-Sachs disease	Hexosaminidase A deficiency
α-Thalassemia	Decreased synthesis of α chain of hemoglobin
β-Thalassemia	Decreased synthesis of β chain of hemoglobin
Wolman disease	Acid lipase deficiency
Xeroderma pigmentosum	UV endonuclease deficiency

From Epstein and Golbus, 1977. *Prenatal Diagnosis of Genetic Diseases.* Reprinted by permission of The Society of Sigma Xi.

Clearly, the biological and genetic changes that can be observed in the populations of industrialized nations are more complex, both in their causation and their consequences, than might first have been suspected. This should provide a note of caution to those who offer superficial remedies for some of our biological and medical problems. Attitudes and values change; the decline in fertility rates seen in the populations of industrialized nations is by no means certain to continue; any radical, especially short-lived, change in fertility patterns may (as in the previously described case from Romania) have startling social and economic consequences. Members of minority groups in multi-ethnic societies may begin to resist intermixture with members of other groups and may initiate a trend toward genetic diversification of subpopulations belonging to a common political body. Under certain circumstances, the use of prenatal diagnostic procedures combined with elective abortion may actually result in an increase in the frequency of deleterious alleles in the gene pool of succeeding generations. Finally, changes in life expectancies which characterize these populations, changes in mortality patterns and in the major causes of mortality, can all be reversed, at least temporarily, by any of a host of possible developments. Evolution remains effective, influencing the genetic composition of human populations, even among the peoples of industrialized, scientifically and technologically advanced societies.

The vast majority of human populations are not part of industrialized countries, but live outside of Europe, North America, or Japan. These societies have sometimes been described as part of a "Third World" or designated as belonging to "underdeveloped" nations. These terms, in addition to expressing ethnocentric biases of Westerners, overlook great variation, even in economic measures, demographic indices, or any of a number of other attributes upon which such classifications are made. Implicit in them, too, is the notion that all societies, in time, will achieve the essential characteristics of modern industrialized nations. Yet, as we have already seen in the case of demographic transition theory, no inevitable outcome follows from similar changes in certain conditions when these occur at different time periods and affect human groups living in a variety of ecological settings. Thus we cannot describe microevolution in human populations solely by reference to the biological and genetic characteristics seen today in the populations of industrialized countries.

Many human populations occupy regions which lack the resources to support industrialization. To take but one example, the many small islands of the Pacific have neither the raw materials nor sufficient numbers of people to support any one of various kinds of industries, while the costs of transportation to and from such areas effectively prohibit the development of manufacturing enterprises. Under colonial control, many islands became suppliers of coconut products (hardly a critical material in the world's economy), of a few nonessential crops such as pineapple or vanilla, or of crops such as sugar which require government subsidies to remain competitive in national or world markets. In the case of only a few islands with rich phosphate deposits from millennia of bird droppings, some native island populations saw their island lands stripped by commercial and governmental projects. By now most of these societies have been exposed to the influences of foreigners for hundreds of years, and traditional patterns of life have been irreversibly altered by the acquisition of new wants and desires, no less than by the demands of foreign administrations. Many of these populations were decimated at first by the introduction of diseases to which they had no resistance, but medical aid, as well as some degree of natural selection, has reversed this process. Even if fertility rates had remained unchanged, these populations would have increased in size as more individuals survived long enough to reproduce and more newly born babies survived, even as mothers began to adopt bottle-feeding and venereal diseases came under control, thus further enhancing fertility.

As in many of the countries which do have the necessary resources and conditions that could support industrialization, these changes in the Pacific area have resulted in shifts of the age distribution of mortality among young children. Diarrheal diseases, which may interact synergistically with such "mild" childhood diseases as chicken pox, are now observed with greater frequency in very young children. Whereas the age of highest risk was once seen in children in the age group of two to three years, occurring around the time when breast milk was replaced by solid foods, diarrheal diseases now appear in children being placed on bottle feeding at only a few months of age. The motivations for the shift to bottle feeding are multiple, including increased opportunities for wage employment in some of these areas. But attitudes about the "modernity," the intrinsic value of a practice observed among the members of a dominant group, also play a role in the rapid spread of the practice. Unfortunately, formulas are often prepared under unsanitary hygienic conditions, the opened cans of milk and prepared bottles remaining unrefrigerated, and the ratio of milk to sometimes contaminated water varies according to the family's economic situation. Babies denied breast milk also fail to receive the antibodies present in mother's milk, so the practice has considerable effects on the nourishment and health of the children, as well as on the fertility of mothers. It is, at best, questionable whether the nutrition of children in these societies has improved with Westernization.

Although the introduction of modern medical programs has led to a decline in the frequency of some diseases, particularly those against which inoculation programs are effective, new kinds of disease pressures affect the health of the population. The introduction of packaged foods has contributed to the accumulation of waste materials, which provide breeding grounds for mosquitos and flies, and to the contamination of local water wells and cisterns. Insect-born diseases have become more prevalent in areas which cannot or do not maintain intensive insect-control programs. The growth of urban or port centers has brought larger numbers of people together in areas lacking facilities for human-waste disposal, and extended families of many members often occupy inadequate housing. New kinds of selection pressures are operating on populations which have only just begun to emerge from a period of decimation from infectious diseases.

In some of these regions, tourism has appeared as the single major industry with immediate potential (Finney and Watson, 1977). But the consequences of tourism–biological, as well as social and cultural–have disturbing effects. The necessary capital for development has usually had to come from outside sources, with the native population limited (for a while at least) to unskilled and low-wage employment. Gene flow between foreign visitors and the native population is likely to increase while social, even ethnic, differentiation proceeds. Until the lengthy process of local capital accumulation and training for skilled and managerial employment is completed, living conditions among the native population suffer in developing urban centers. The tourist industry is fragile, acutely sensitive to economic recessions in other parts of the world; at most, only relatively few members of the native population are likely to participate fully in the material benefits of these developments. Social class differences—extending to differential access to medical care, mate selection, diet, and other aspects of the social and economic system—are probable, if not inevitable, outcomes of this limited form of economic development.

This somewhat dismal picture is scarcely the invariable outcome of limited integration into the world economy by populations living in areas unsuited to industrialization. Careful planning could obviate many of these problems; in a few areas, native populations are resisting outside pressures for rapid development. But, with rare exceptions, most indigenous peoples cannot themselves accumulate the capital to control industry; the breakdown of traditional political systems often puts the control of development into the arena of foreign manipulation of weak native factions. From this viewpoint, it seems unlikely that in the near future the populations of such areas will exhibit certain of the biological changes which have been noticed in the peoples of industrialized nations.

First of all, it does not seem likely that the secular trend toward increased stature and decreased age at menarche will be evident, or so strongly marked, among these populations. Dietary changes are taking place, but it is by no means clear that these actually represent improvements in nutrition. Rather, as native food plants are ignored in favor of costly foreign imports, the quality and quantity of intake for wage earners and their families may actually be worsened, except among those few who form a native elite in the new economy. One of the striking changes I have observed in dietary practices in some island populations, and the same has been reported elsewhere by others, is a decline in the consumption of fish, the single major source of protein on small islands. While the spread of outboard motors for boats is extensive, fishing has become less intensive in many areas; the two major effects of the introduction and spread of outboard motors have probably been an increased rate of travel and communication between people living on different islands or on different parts of larger islands and a decreased rate of loss of life at sea. Moreover, as native populations increase in numbers, one effect may be to reduce individual caloric and protein intake, thus impairing growth processes. The major advantage of this dietary trend would be its possible effect in suppressing fecundity and maintaining or extending the period of adolescent sterility. Even this, however, would have slight impact on fertility rates, in contrast to the effects of relaxing traditional sexual tabus, the reduction of venereal diseases, and the adoption of bottle feeding.

Many of these societies have been strongly affected by missionary influences and some still endorse values they have learned in this manner, values which discountenance contraceptive practices. Even when prevailing ethical and moral values are not opposed to the adoption of contraception, the most reliable methods are the product of Western technology and must be imported, making their supply both expensive and uncertain. At the same time, some of the older, native practices which limited fertility have been discarded. In a number of the island populations with which I am familiar, men formerly spent a considerable amount of time in men's houses, isolated from women, and restraints on sexual activity prior and after fishing expeditions were common. These restrictions are no longer practiced, and others, such as lengthy postpartum abstinence, are falling into disuse. And, of course, intentional infanticide is now rarely reported, although differential care ("indifferent infanticide") of children may have some comparable effect on infant and child mortality rates.

Many of these societies cannot support the medical facilities which have been made available to them by foreigners, yet none is now likely to willingly forego the modern medical care to which they have been exposed. Either a disproportionate amount of the limited income available to them must be diverted to this source, thus further limiting their ability to accrue investment capital, or they must continue to depend on external economic aid–either as a form of charitable economic dependency or in return for concessions to outside interests seeking military bases, oil superports, or commercial control over tourist or fishing industries. Whatever arrangements are made, the results of continued access to modern medical advantages are likely to accentuate population growth, unless family planning programs are successfully instituted and widely adopted.

In any case, genetic differentiation of localized populations is decreasing as mobility increases, as traditional mating practices are discarded, and as new opportunities for social and economic advancement appear. Emigration is becoming widespread, especially among young adults, although the conditions of living among the migrant groups in their new homes are rarely an improvement on conditions in the home island. The loss of these young adults represents a form of continuing genetic drift; and since many migrant groups remain economically, culturally, and biologically isolated from their new neighbors for several generations after their arrival, unique gene pools are being repeatedly formed and tested under new environmental conditions.

The biological and genetic changes which have accompanied industrialization in some countries may appear in the populations of other countries that have the resources capable of supporting a similar pattern of development, but there is no reason to expect that a kind of unilineal evolution will take place in all the industrializing societies of the world. Even in the most highly industrialized nations, segments of the national population continue to live under appalling circumstances, lacking adequate nutrition, housing, education, or access to medical care. These groups, sometimes described as belonging to a "Fourth World," comprise a population subject to unique biological pressures and characterized by distinctive demographic indices of biological fitness. If the members of such groups are not more fully integrated into the larger society of which they are part, with avenues of individual mobility opened or expanded for them, even the highly industrialized societies will be characterized by distinctive biological sub-populations.

In some of the so-called developing countries, class differences may be even more formidable barriers to common evolutionary trends. In some of the countries which have adequate natural resources for industrialization, capital has historically been concentrated in the hands of a small elite class, many of whom have been reluctant to invest their wealth in the economic development of their own nations. Economic development in these countries has too often been based on capital investment from outside sources, with profits being returned to foreign investors, and with managerial and professional employment opportunities being exploited by foreigners or by members of the small elite native group. Most of the natives in such countries have too often had to accept low-wage employment or sharecrop farming without benefit of social legislation to ameliorate living and working conditions and without hope of any greater degree of participation in the benefits of economic development. Even when political movements have sought to redress these inequities, the absence of trained and educated personnel or of investment capital has hindered the implementation of the noblest ideals.

In many of these countries, genetic diversity among sub-populations and between classes appears commonplace. Ethnic and racial groups remain distinctive entities despite formal governmental pronouncements endorsing equality, and socioeconomic differences restrict access to goods and services. Even when racial or ethnic membership is held constant, certain phenotypic differences exist between and among classes in such fundamental biological features as the growth rates of children and the body size of adults. As in the United States (where even so sensitive an index as infant mortality rates varies greatly for different racial and ethnic groups), mortality rates, life expectancy values, and other demographic indices in these developing nations differ by social class and by ethnic or racial group. Thus, regional differences may be receding as mobility changes and urbanization take place, but socioeconomic differences have become effective barriers to gene flow and to an equitable distribution of resources, contributing to microenvironmental differences.

If all societies undergo similar changes as industrialization proceeds, then it might be possible to predict that the biological characteristics of human populations in industrializing societies are merely representative of a universal stage in human evolution. But it is now far less certain that the Euroamerican experience is a sufficient model for studying industrialization. Rather, industrialization may follow other courses, of which the Russian and, particularly, the Chinese cases stand as examples. The biological and genetic consequences of the Chinese alternative path to industrialization are simply unknown, although it is certain that the political changes which have taken place were accompanied by great turmoil, population movements, and, in both Russia and China, by considerable loss of life. Yet, even in the absence of hard data, it is at least reasonable to expect that the populations of these countries have undergone genetic changes unlike those of other industrializing countries.

New ideals, new wants and aspirations, and more widely accepted attitudes about human relationships and responsibilities are playing a significant role in the evolution of human societies and nations. These must affect human evolution through indirect consequences of changes associated with economic development, the increased rate and intensity of interaction between individuals and nations, and the greater social and economic interdependence of peoples and nations who dwell on a planet of finite resources. But we also share in a world of ideas and knowledge, available to all who have access to educational opportunities, and through these we can affect biological evolution to an extent never before possible. This ability to consciously influence biological processes imposes a heavy responsibility on us all. It does not end human evolution, but extends the evolutionary processes to levels of complexity with which the animal geneticist need never deal.

REFERENCES AND RECOMMENDED READINGS

CRUZ-COKE, R. 1970. *Color Blindness: an Evolutionary Approach* Springfield: Charles C. Thomas.

EPSTEIN, C.J., and M.S. GOLBUS 1977. Prenatal diagnosis of genetic diseases. *American Scientist,* 65:703–711.

FINNEY, B.R., and K.A. WATSON, eds. 1977. *A New Kind of Sugar: Tourism in the Pacific.* Santa Cruz: Center for South Pacific Studies, University of California at Santa Cruz.

HARTUNG, J., and P. ELLISON 1977. A eugenic effect of medical care. *Social Biology,* 24:192–199.

KOLATA, G.B. 1977. Research News: Obesity, a growing problem. *Science,* 198:205–206.

MARQUARDT, H.; F. RUFINO; and J.H. WEISBURGER 1977. Mutagenic activity of nitrite-treated foods: human stomach cancer may be related to dietary factors. *Science* 196:1000–1001.

RETHERFORD, R.D. 1974. Tobacco smoking and sex ratios in the United States. *Social Biology* 21:28–30.

RICHARDSON, W.N., and T.H. STUBBS 1976. *Evolution, Human Ecology, and Society.* New York: Macmillan Publishing Co.

SKLAR, J., and B. BERKOV 1975. The American birth rate: evidences of a coming rise. *Science*, 189:693–700.

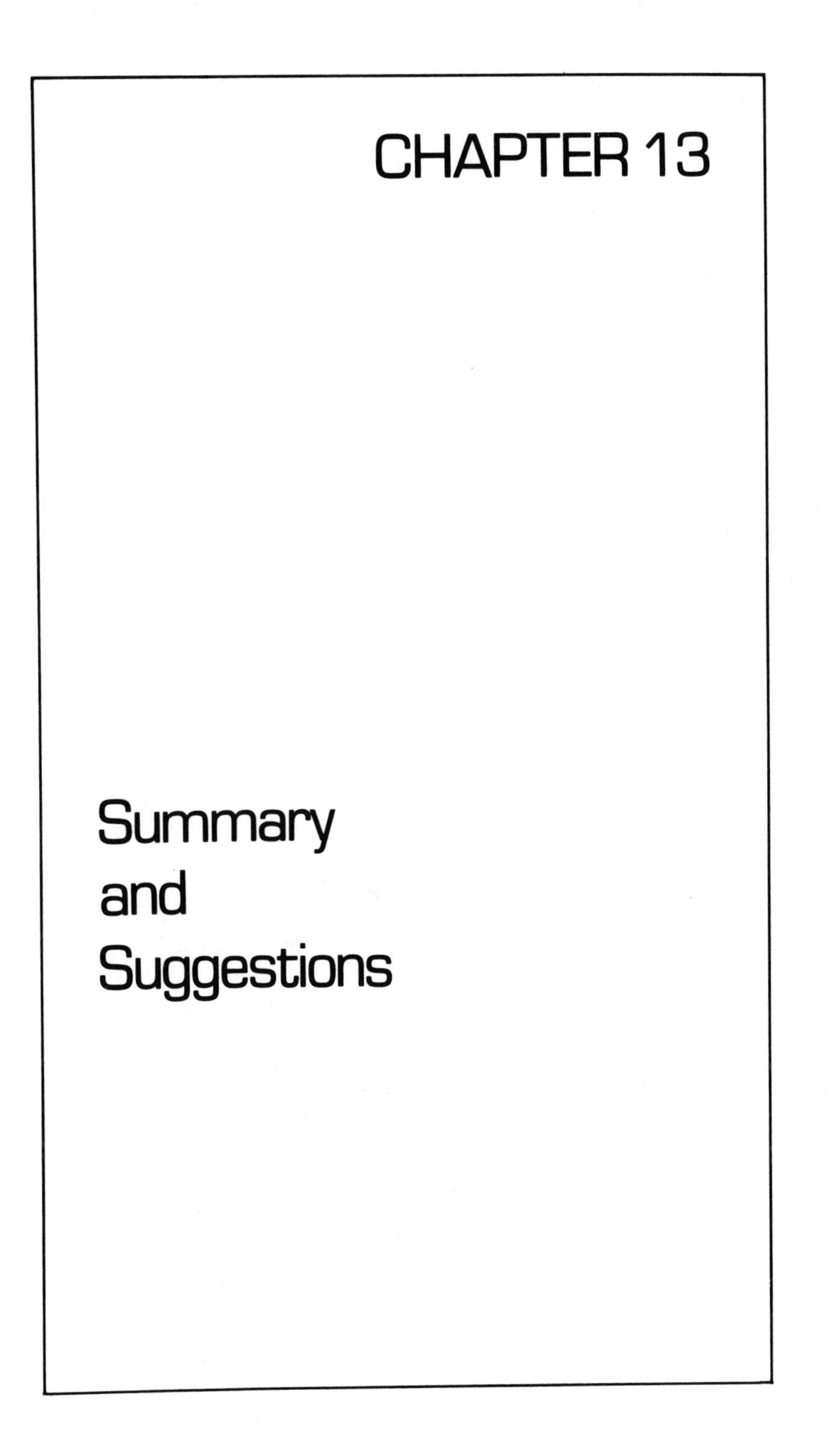

CHAPTER 13

Summary and Suggestions

It has been a long path from the initial discussion of the use of palmar creases in medical diagnoses to comments on the biological and genetic consequences of Maoism, but the subject of human variation and evolution is even more complex than the range of topics touched upon in this book can begin to suggest. Because basic knowledge about human genetic diversity has come from geneticists, the fundamental principles of human biochemical genetics are derived from studies of the structure and function of the genetic materials, the nucleic acids, in other organisms, and mainly from studies of bacteria, molds, and other small animal forms. We have also borrowed extensively from the models and principles developed from studies of other animals to study the population genetics of human groups. Because we are part of the organic world, part of what was once called the "animal kingdom," we are subject to biological processes, ourselves the product of organic evolution, no less than are our animal cousins.

But other biologists cannot provide us with the methods or models which we need to fully comprehend the study of human evolution; our species is the product of biocultural evolution, and students of human diversity must consider the effects of technology and ideology, of religion and social organization, of shared values, attitudes, ideals and knowledge on the biological and genetic heritage of human populations. The task is formidable and challenging; the results, however imperfect yet, will be awesome. While a comprehensive understanding still lies ahead, it may be worthwhile to attempt to summarize some of what has already been accomplished.

In previous chapters, each of the major evolutionary processes—genetic drift, mutation, gene flow, natural selection—has been examined in artificial semi-isolation from the others. This is a useful procedure for developing some degree of familiarity with each source of genetic change, but one which is remarkably divorced from reality. In a population which to some degree is engaging in assortative mating, a practice common to human groups, mutations are occurring at the same time that selection is taking place; migrants are entering and leaving the group; and some couples, independent of genotype differences or similarities, are producing more children than others. Many published studies in human micro-evolution tend to focus on the operation of only one of these evolutionary forces affecting a study population. This is, at best, a practical device, allowing the investigator to emphasize a single subject—without denying that the genetic diversity observed in any single population is the product of the interaction of several evolutionary forces, which vary in the relative degree of their influence on the gene pool of the population.

We know, for example, that the efficiency of selection depends on the mode of inheritance of the trait that confers on its possessor a selective disadvantage. If a lethal condition, one causing death before or shortly after birth, is caused by a dominant allele, every time the allele appears and is expressed in the phenotype of an individual, whether homozygous or heterozygous at this locus, it is immediately removed from the gene pool. The reappearance of the allele under these conditions depends wholly on the mutation rate v ($a \rightarrow A$), since the coefficient of selection is 1.00 against both the homozygous and heterozygous genotypes. It follows that where $s = 1.00$, the frequency of the dominant allele in the gene pool is equal to the mutation rate. Thus, if $s = 1.00$ and $v = 0.10$, the equilibrium frequency of the allele p will be 0.10. More often, of course, it has a value less than 1.00, and the frequency of the allele at equilibrium will be a function of the relationship between v and s, expressed in the formula, $p = v/s$.

However, if a disadvantageous trait is expressed in individuals with the homozygous recessive genotype, the allele is subject to removal from the gene pool only when it appears in the homozygous genotype combination, and the recessive allele is protected from selection when it is present in the heterozygote. The frequency with which the homozygous recessive genotype will appear depends, of course, on the frequency of the recessive allele in the gene pool. Thus, an allele with a frequency of 0.10 can be expected to appear in homozygous combination in 100 out of 10,000 genotypes, but if the recessive allele frequency is 0.01, only one out of 10,000 genotypes can be expected to have the homozygous recessive combination. Since selection operates against the whole organism, the allele frequency will not be altered by natural selection until the allele reaches frequencies at which homozygous recessive genotype combinations are formed. In the case of lethal phenotypes ($s = 1.00$), the rare allele normally attains this frequency only as a result of the accumulation of mutations which are usually carried, over preceding generations, in the heterozygous condition where they are protected from the effects of natural selection. It follows, then, that the equilibrium frequency of a recessive allele that produces a lethal condition in the homozygous combination is also dependent on the value of both the rate of mutation and the coefficient of selection, but in this case the relationship is expressed as $q = \sqrt{u/s}$. Accordingly, in contrast with phenotypes produced by a dominant allele, the rate of mutation must be much higher in order to produce an equivalent frequency of phenotypes manifested by the homozygous recessive combination against which selection can operate.

The consequences of these relationships have not always been appreciated by those who have argued more vociferously for the need to introduce some forms of eugenic practices, "negative eugenics," to improve the gene pool of human populations. Preventing individuals with deleterious genetic conditions from contributing to the gene pool, either by extermination or sterilization, will not result in the final removal of the responsible alleles from the species. Conditions caused by the presence of an allele involving a dominant mode of inheritance will continue to appear as a result of recurrent mutation; so, negative eugenic procedures would have to continue so long as the species survives. In the case of conditions expressed by individuals with a homozygous recessive genotype, sterilization or elimination of the affected will, at best, reduce the allele frequencies to levels at which the allele is more likely to be present in heterozygotes and, thus, more likely to be protected from manifestation and removal.

It is obvious that every human population carries a burden of deleterious mutations, which impairs the fitness of the group. As defined by Crow (1960), *genetic load* refers to the proportion by which fitness is reduced in the population due to the operation of a factor such as mutation. If 10 percent of a population carries a mutant dominant allele, the mean fitness, $\bar{W}$, would be less than if the mutant allele were not present ($\bar{W} = 1.00$). But the number of "genetic deaths" due to the *mutational load* in a population entails the effects of mutant alleles at more than one locus, and "genetic death" can result from decreases in fertility as well as from increases in mortality and morbidity due to deleterious mutations.

The mean adaptedness of a population may also be reduced by factors other than mutation. Consider, for example, the effects of maintaining a balanced genetic polymorphism through heterozygote advantage, such as that for the sickle hemoglobin allele in West African populations. The enhanced fitness of the heterozygote at this locus ensures the retention of both alleles in the population. Thus, a certain number of both kinds of homozygotes will continue to be formed and to be removed from the population at rates determined by the relative selective disadvantage experienced by each homozygote in the malarial environment. This *segregational load* does not depend on the production of new mutant alleles, but on the consequences of the selective advantage held by heterozygotes who produce gametes of both kinds.

In contrast, an *incompatibility load,* the result of maternal-fetal incompatibilities, arises from certain forms of interactions between the fetus and a mother whose genotype is unlike that of the unborn offspring. As described in earlier chapters, the best known of these interactions involves the production of anti-*Rh* positive (anti-*D*) antibodies by a mother who is homozygous for the recessive allele (*dd*) but is carrying a heterozygote child (*Dd*). Incompatibility interactions also occur, more rarely, in the fetus whose *ABO* antigen is at risk from the appropriate antibody present in the mother's plasma; and other kinds of incompatibilities are also known to exist.

It is useful to consider these and other factors that are capable of reducing the adaptive fitness of a population as comprising the total genetic load. Attempts to distinguish the relative contribution of mutational and segregational loads to the total genetic burden of a population have been repeatedly challenged, and it is particularly difficult to evaluate this measure when applied to human populations. Individual human carriers of a deleterious mutation may experience "genetic death" in the sense of failing to produce offspring, yet may contribute to the adaptive fitness of the population by, for example, altering the environment so as to enhance the mean adaptive fitness of the population. How many reproductively effective children did Madame Curie have, yet how many victims of cancer have survived and subsequently produced offspring because of therapeutic treatment with radium, the element which she discovered and studied? We can increase the relative contribution of the mutational load to the total genetic load present in the population of Hiroshima by releasing an atomic bomb over the city, or we can produce an increased frequency of deleterious homozygous recessive genotypes against which selection can operate through inbreeding. The point to be made here is that the creation and maintenance of human genetic diversity results from the interaction of evolutionary forces operating on a culture-bearing species.

The reproductive behavior and history of fruit fly populations maintained over many generations in closed vials can be directly observed by the laboratory geneticist, but the student of human populations must collect genealogies, identify the diseases which have affected human groups, and record the reproductive histories of natives who may lack either a system of writing or a tradition of record keeping. The presence of an allele which is rare in neighboring groups but found in high frequency in the population being studied may not be due to an unusually high rate of mutation or to the operation of some unique form of selective agency, but only to the reproductive overachievement of an itinerant visitor or to the unrepresentative genetic composition of a few founding ancestors. A seemingly homogeneous human group may be erroneously described in terms of allele frequencies that are merely the average of summed values distributed among several Mendelian populations separated by differences in a complex of beliefs, values, and practices which are symbolized by whether one kneels or stands to pray to a deity in whom all believe. It takes little effort to recount a number of ways in which cultural factors can affect the processes of human evolution or to demonstrate how critical anthropological data are to the study of human population genetics.

But much of the cultural information which the investigator must acquire depends on incomplete or imperfect recollection, and some data are simply not known by native informants. The records of early foreign explorers may indicate that the disease which killed so many people, described by natives as the result of witchcraft, was really smallpox. And this fact, in turn, may help to explain why a smallpox epidemic did not occur among the survivors when a stricken smallpox victim was left among this people a few years thereafter. The church records of births and deaths kept by dedicatea missionaries may augment the failing memory of an elderly informant living in a society which forbids the use of a name formerly belonging to someone who has died. In short, the techniques and methodology of ethnohistory—the use of oral and written materials to elicit the processes of culture change and culture history—provide invaluable assets to the student of human diversity. This approach has, for example, been highly productive in attempting to analyze the components of selection affecting the genetic composition of native populations.

Selection may operate through differential fertility or differential mortality; and given the different conditions under which human societies exist, populations vary in the relative contributions made by these two components to the total selection intensity to which each population is exposed. Crow (1958) formulated such an index of total selection intensity, (I_t), which is composed of two components: I_f, due to fertility differences among women who have attained reproductive age; and I_m, due to mortality differences among females during the period between birth and attaining reproductive age. For example, in a population with high rates of mortality among infants and young children, only a small proportion of female infants would survive to the age at which marriage and reproduction begin. Those females who do survive long enough to reproduce may give birth to variable numbers of children, but selection has probably been more intense in terms of the differential mortality of females during the pre-reproductive years. In contrast, where infant and childhood mortality rates are very low, selection can be expected to operate relatively more severely on fertility differentials of women who have attained the age at which reproduction can begin.

Obviously, mortality and fertility rates are not determined solely by genetic factors, but by a complex of biological, environmental, and cultural variables. In a society which makes advanced medical treatment and adequate nutrition available to all citizens, relatively few children are likely to die from infant or childhood diseases, and variations in fertility would contribute to a relatively greater degree to the total selection intensity. In other societies, inadequate nutrition accompanied by high rates of infant diarrhea, even in the absence of communicable diseases among the young, may result in high rates of child mortality and the relatively greater contribution of mortality differentials to the total selection intensity. It is more reasonable, then, to refer to these components as comprising an index of the *opportunity* for natural selection.

The value of an analysis of these components, measured indirectly from demographic data or the vital statistics of a population, is that comparisons can be made among populations, between segments of a population, or of changing patterns of the relative contributions of fertility and mortality factors in a single population over time. In the United States over the last hundred years, mortality differentials have contributed relatively less to the total index in recent decades than in those earlier periods of our history when, for example, diphtheria ranked as one of the ten leading causes of death. Within American society, this change over time to the relatively greater opportunity of fertility differentials to contribute to selection has been uneven—first affecting affluent segments of the populace and Euroamericans, and appearing only later, if at all, among the economically deprived and racial minorities.

The opportunity-for-selection index has been calculated for a number of populations, and comparisons suggest that a trend exists toward a decrease in the mortality component of selection opportunity in technologically advanced societies. Despite the trend in such societies toward the wide use of reliable contraceptive techniques, variation in fertility continues to occur, making the contribution of this component relatively greater. Cultural factors, our ability in various ways to modify the environment and alter the selective forces affecting the gene pools of human populations, introduce high levels of indeterminism into the course of human evolution and virtually guarantee the continuing opportunity for the operation of natural selection on human populations.

But if selection continues to operate primarily and to an increasing degree in terms of fertility differentials, as seems likely within many contemporary human populations, what can be said about the influence of other evolutionary processes? Increased mobility within and among human societies and countries probably spells an end to the kind of regional isolation which once led to the growth of highly differentiated populations. But this scarcely implies complete homogenization of differences in a species that erects barriers to intercourse based on political, economic, religious, and ideological differences. Never again, however, are human groups likely to remain reproductively isolated for such lengthy periods as they once were, for the reproductive barriers between contemporary populations are essentially cultural, not geographic. Genetic diversification is likely to continue, with new groups forming and dissolving as socioeconomic differences appear and change, as new religious sects form and evolve into more traditional and ecumenical churches, and as political movements arise and spread within and among peoples.

If the role of genetic drift were not sufficient to maintain genetic diversity in the species, mutation would surely continue to contribute to human variation as technological changes take place. The use of radioactive materials in medical diagnoses and treatment, in industrial development, and in the development of energy supplies, assures that mutagenic agents will continue to be produced. And the possibility that humans may begin a period of extended exploration of space in the near future adds a new potential source of mutation to our environment. Our growing ability to maintain life and to extend the lifespan and productivity of those who carry mutant genotypes will also contribute to the retention of genetic diversity resulting from mutation.

The diversity of human populations, our guarantee against extinction as a species in the changing environments of future millennia, is not about to disappear. If humanity is to survive, we have no choice but to appreciate that diversity, to understand the effects of environmental alterations on our genetic heritage, and to evaluate the possible effects of our actions and attitudes before we undertake or permit changes affecting the adaptations of human populations, present and future, to their environments.

REFERENCES AND RECOMMENDED READINGS

CROW, J.F. 1960. Mutation and selective balance as factors influencing population fitness. *Molecular Genetics and Human Disease.* L.I. Gardner, ed. Springfield: Charles C. Thomas.

——. 1958. Some possibilities for measuring selection intensities in man. *Human Biology* 30:1–13.

DOBZHANSKY, T. 1970. *Genetics of the Evolutionary Process.* New York: Columbia University Press.

MORGAN, K. 1973. Historical demography of a Navajo community. In *Methods and Theories of Anthropological Genetics.* M.H. Crawford and P.L. Workman, eds. Albuquerque: University of New Mexico Press. Pp. 263–314.

UNDERWOOD, J.H. 1975. *Biocultural Interactions and Human Variation.* Dubuque: Wm. C. Brown.

APPENDIX

Anthropological Studies and Biochemical Variation

While anthropological studies of human variation traditionally concentrated on measuring head shape, nose form, stature, and similar traits, most of which involve polygenic inheritance, the discovery of the *ABO* blood-group differences in humans by Landsteiner in 1900 opened up the field of studies of biochemical variation in and among human populations. Since most individuals are unaware of their own blood-group classification, matings are not based on these features; as we learned more about the various blood group systems, it appeared that most, if not all, followed a pattern of simple Mendelian inheritance, while the blood group substances themselves involved a relatively direct expression of the responsible genes. It seemed likely that these traits were immune from modification by dietary influences, disease history, or by any of a host of factors which were already known or suspected to affect the phenotypic expression of the anthropometric traits that physical anthropologists had been investigating for so many years. Beginning in the first half of this century, physical anthropologists started collecting blood samples from members of the populations which they studied, and an increasing emphasis was placed on the use of information from blood-group studies to unravel problems of human microevolution.

The blood group systems we investigate are usually designated in terms of the antigenic substances present on the surface of circulating red blood cells, but also involve the antibodies to these antigens which are found in the liquid portion of the blood, the *blood plasma.* As we now know, the antigens of the best-known blood-group system, the *ABO* blood groups, are complex macromolecules formed from precursor substances in the red blood cell, known as glycolipids, combinations of carbohydrates and fatty acids. *O* blood type, for example, results when a precursor substance is transformed into another substance in the presence of the Le^a allele. This latter material, which we can term the Le^a substance, is modified by an enzyme produced by the *H* allele to add a sugar at a specific site on the carbohydrate chain of the Le^a substance. Since the resulting "*H* substance" is not further modified by the *O* allele, it might be more accurate to refer to this system as the *ABH* blood group. But old terms have become imbedded in the language, and we can avoid confusion by referring hereafter to the *ABO(H)* blood-group system. In the presence of the *A* or *B* allele, other enzymes add specific sugars at different positions of the *H* substance to produce structurally distinctive products, the *A* antigen or the *B* antigen.

The term *antigen* is derived from combining parts of two words, *anti*body *gen*erating, and these antigenic substances interact with appropriate antibodies when brought together. *Antibodies* are specific kinds of proteins, called immunoglobulins, released into the blood plasma when they are produced by specialized antibody-forming cells in response to the presence of foreign antigenic substances. If red blood cells from an individual with *A* antigen are mixed with serum taken from a person with *B* blood type, the red blood cells will usually *agglutinate,* or clump together. This interaction can be thought of in terms of a lock-and-key matching, involving the binding together of matching molecular configurations of the *A* antigenic substance and the anti-*A* antibody regularly present in the plasma of individuals who have the *B* antigen on their red blood cells.

Some blood group systems are readily identified by mixing washed red blood cells with serum containing the specific antibody and observing the clumping together or agglutination of the tested cells exposed to this antiserum. But other blood group system interactions require more complex testing procedures and the addition of other materials before distinctive antigen-antibody reactions take place. The *ABO(H)* blood group system interactions involve what are sometimes termed "naturally occurring" antibodies, which can usually be detected in the plasma shortly after birth. But some other blood group systems, such as the *MN* system, are not usually represented by antibodies circulating in the blood stream, and anti-*M* and anti-*N* antisera are most commonly obtained by immunizing rabbits with human erythrocytes.

The basic features of the *ABO(H)* blood-group system can be described in terms of three alleles which can form any one of six possible genotype combinations, *AA, AO, BB, BO, AB,* or *OO,* to produce four distinguishable antigenic phenotypes, *A, B, AB* and *O.* These four phenotypes can be distinguished by testing red blood cells with only two antisera, anti-*A* and anti-*B* antisera. Obviously, a person whose red blood cells bear the *A* antigen could not have anti-*A* antibody in the plasma or agglutinated clumps of red blood cells would block the blood vessels or lodge in the blood vessels supplying critical organs such as the heart or brain; so, it is possible to assume the characteristic antibody phenotype of individuals for whom the *ABO(H)* antigen identification is known (Table A). It follows from these relationships that antibody-antigen interactions, even though some of these may be weak responses, are possible between the red cells of an individual of any antigenic category and the antibodies present in the plasma of individuals belonging to one of the incompatible blood types.

Despite comments to the contrary found in some older textbooks, transfusions from donors belonging to a different *ABO(H)* blood-type category always involve some degree of risk. There are no "universal donors" or "universal recipients," and blood transfusions must always be preceded by cross-matching, a test of interaction between the recipient's red blood cells and the intended donor's plasma before transfusion begins, even when both have identical *ABO(H)* blood types. In part, this procedure guards against incompatibilities for blood group systems other than the

TABLE A Outline of Basic ABO(H) Blood Group System Identification Procedures

Antisera	*Reaction*	*Antigen Phenotype*	*Antigen Genotype*	*Antibody Phenotype*
anti-A	positive	A	AA, AO	anti-B (anti-H)*
anti-B	negative			
anti-A	negative	B	BB, BO	anti-A (anti-H)*
anti-B	positive			
anti-A	positive	AB	AB	(anti-H)*
anti-B	positive			
anti-A	negative	O	OO	anti-A anti-B
anti-B	negative			

*May be present

ABO(H) one, but it also ensures that transfusions do not take place between individuals who have anomalous phenotypic expressions of the alleles present at this locus, such as the Bombay phenotype to be described below.

In fact, the inheritance and expression of alleles in the *ABO(H)* blood-group system are more complex than has been indicated thus far. For one thing, refined techniques have established the existence of several variants of the *A* antigen, the more common of which are those produced by the alleles A_1 and A_2; so the number of possible distinctive phenotypes for this blood system is much greater than six. It should also be recalled that the *H* precursor substance is produced independently of the *ABO* locus, and the presence of the *A* or *B* alleles at the latter locus results in the formation of enzymes which convert the *H* substance to *A* or *B* antigens. Thus, at least two loci interact in the production of antigens in the *ABO(H)* system, but in most cases the antigen found on the red blood cells can be described in terms of a simple mode of single-trait Mendelian inheritance determined by alleles at a single locus. However, rare individuals are homozygous for the *h* allele, and these persons do not form *A* or *B* antigens even when the genotype includes the appropriate allele (Bombay phenotype).

In a classic example of the interaction of these two loci, an *O* woman married to an *A* man gave birth to a child with *AB* blood type. The birth of an *AB* child to a woman with *O* blood type cannot satisfactorily be explained as a result of illegitimacy; when the woman's blood was tested further, she was found to have not only the anti-*A* and anti-*B* antibodies, but also anti-*H* antibodies which are not found in typical *O* persons. The woman's father was a typical *O,* and her mother was *B* blood type. Further studies revealed that the genotype of the woman was *BOhh* and that of her husband *A-H-*. Their son, having received an *H* from his father, expressed the *A* allele from his father and the *B* allele received from his mother, but the *B* allele present in the woman's genotype was not expressed on her red blood cells in the absence of the *H* allele.

The interaction between the *ABO(H)* locus and another locus, secretor, actually facilitates certain kinds of blood type identifications. In some populations, such as the Australian aborigines, the *A* and *B* antigens can usually be detected in saliva and other mucus secretions, since most members of these populations have a dominant allele present at the secretor locus. This allele directs the production of a water-soluble form of the *A, B,* or *H* antigens in various bodily secretions. The frequency of the dominant secretor allele varies among populations, however, and saliva samples cannot substitute for the blood samples which anthropologists must collect, often with great difficulty, from the human populations which they study. Finally, it should be noted that at least one other genetic locus interacts with the *ABO(H)* and secretor loci in the expression of the *ABO(H)* antigens in red blood cells and bodily secretions. The Lewis antigens, Le^a and Le^b, are water-soluble antigens, and the interactions between this locus, the secretor locus, and the *h* locus are relatively complex.

In addition to the *ABO(H)* blood-group system, a number of other antigenic systems have also been discovered and are used to study human variation. For nearly twenty years after the discovery of the *MN* system in 1927, it was thought that two alleles at this locus, *M* and *N,* produced three distinguishable phenotypes, *M, MN,* and *N,* detected by two antisera, anti-*M* and anti-*N.* Unless sensitized by prior blood transfusions, antibodies to these antigens are not usually present in the serum, and the antigens are of limited medical interest in contrast with certain other systems, particularly the *ABO(H)* and *Rh* blood groups. But the discovery of anti-*S* antiserum in 1947 quickly revealed that another antigen, *S,* is commonly found in association with the *M* antigen more often than should occur by chance in a number of tested populations, and studies of pedigrees in some of these groups suggest that *M* and *S* alleles segregate as a pair. Subsequently, a fourth antiserum, anti-*U,* was identified, which may represent an allele at the *S* locus (S^u) that seems to suppress the expression of the *S* antigen. Other complexities affecting expression of the *MNS* system have now been identified, including the Henshaw antigen which in some West African populations seems to have a high degree of association with *NS.*

The *Rh* (Rhesus) blood group system was originally detected in 1940, with antisera obtained by immunizing rabbits with red blood cells from rhesus monkeys. It was thought at first to involve a single genetic locus with two alleles, *Rh* positive and *Rh* negative, producing two phenotypes, Rh^+ (with the *Rh*-positive allele acting as a dominant allele) and Rh^-, produced by the homozygous recessive *Rh*-negative genotype. This blood group became of considerable medical interest when it was recognized that cases of *erythroblastosis fetalis,* a severe form of jaundice in the newborn, could occur when the mother was Rh^- and the child was Rh^+. Subsequent discoveries revealed that a larger number of *Rh* antigens exist than this simple biallelic system can account for, and additional antisera were developed from human plasma which allow several kinds of *Rh* antigens to be detected. The complexities of this system have been described in terms of two genetic models.

Fisher's system proposes that the *Rh* antigens are produced by allelic-pair combinations at three closely linked loci, *C, D* and *E,* with *D* allele representing what had earlier been termed the *Rh*-positive allele and *d* allele the *Rh*-negative allele. In the more common form of *Rh* incompatibility interaction of medical concern, the genotype of the Rh^- mother would have been *dd,* and that of the Rh^+ child *Dd.* But, since three closely linked loci are involved, eight kinds of gametes, or haplotype allele combinations, at this locus could be formed, as indicated in Table B.

An individual's genotype would be represented, of course, by combinations of two of these haplotypes. While five antisera are available for detecting these combinations, no anti-*d* antisera has yet been found; so it is still impossible to distinguish the heterozygote *Dd* from the homozygote for *D,* while the homozygote for *d* is distinguishable only on the basis of the absence of any reaction between a blood sample from such a person and anti-*D* antisera. Further, while the available antisera permit us, for example, to detect the presence of a genotype combination such as *CcddEe,* it is not possible to specify the precise haplotype combinations which, in this case, might be *CdE/cde, Cde/cdE,* etc.

Wiener, however, proposed that the various *Rh* antigens were the product of combinations of pairs of at least eight possible alleles at a single locus, as indicated also in Table B. This system of notation has the advantage of simplicity, but it is not yet certain which model of inheritance is more appropriate. Several additional variants of the more well-known antigens of the *Rh* system have also been discovered, including a set of antigens which react weakly with anti-*D* antisera (D^u variants) and others which react with a more recently identified anti-*V* antisera. In a few cases, parents with detectable *D* antigens have produced one or more children whose blood does not react with anti-*D* antisera—suggesting that, as in the case of the *ABO(H)* system Bombay phenotype, other genetic loci are involved in the expression of the *Rh* antigens.

In addition to these three more well known blood-group systems, a dozen or more other antigenic systems have been identified and used in studies of human blood-group polymorphisms. These include the Diego antigen (Di^a) which appears to be inherited as a dominant; the Xg^a antigen which is determined by an *X*-linked allele; the Duffy system produced by the alleles Fy^a and Fy^b to produce three antisera-reactive phenotypes and the negative $Fy(a^-)Fy(b^-)$, which does not react with any available antisera; the Lutheran system, also with three phenotypes and rare negative-

TABLE B Major Kinds of *Rh* Haplotypes, Showing Comparison of Fisher-Race and Wiener Systems of Notation

	Haplotypes						
Fisher-Race notation	CDE	CDe	cDE	cde	Cde	cdE	CdE
Wiener notation	R_z	R_1	R_2	r	r′	r″	r_y

reactors; and a number of other systems of varying degrees of genetic complexity. With some fifteen or twenty blood-group systems available, blood typing tests became useful tools in resolving legal cases of disputed paternity.

The results of blood typing comparisons cannot be used to prove paternity, but it is possible that an individual can sometimes be reasonably excluded as the biological father of a child whose paternity is being contested. For example, suppose a mother whose blood types were *A*, *Rh*-negative, and *MN* claimed that her child, whose blood types were *AB*, *Rh*-positive and *M*, was the son of a man who was shown to have blood types *O*, *Rh* positive, and *M*. Obviously, a man with *O* blood type cannot father an *AB* child, although the findings are consistent with the claim at the *Rh* and *MN* loci. The results of an exclusion test can, of course, be strengthened by consideration of a larger number of antigenic systems. In the last few years, however, the use of red blood cell antigenic systems in paternity cases has been declining in favor of tests of similarities and differences in the histocompatibility (*HL-A*) system, antigens present in most tissues other than the red blood cells, which have proven more sensitive and discriminating.

Indeed, while studies of red blood cell antigens are perhaps the most common method of studying biochemical polymorphisms in human populations, a growing number of other biochemical traits has proven of great value in anthropological research. The traits used in such research include studies of variant forms of hemoglobin, of such enzymes and their variants as glucose-6-phosphate-dehydrogenase (G6PD) and red-cell acid phosphatases, and of the distribution of variant forms of several kinds of serum proteins such as the haptoglobins and transferrins, and two kinds of immunoglobulins, the *Gm* and *Inv* factors.

Almost 95 percent of the protein in circulating red blood cells is hemoglobin, a conjugated protein containing an iron complex of poryphyrin, *heme*, which has a strong affinity to oxygen. The protein portion of hemoglobin consists of amino acids joined by peptide bonds to form two paired polypeptide chains. Different paired chains most commonly found in adult hemoglobin are designated as $\alpha_2, \beta_2, \gamma_2$, or δ_2, so that the notation for normal adult hemoglobin *A* is $Hb_{\alpha_2}{}^{A}{}_{\beta_2}{}^{A}$, indicating that the hemoglobin *A* molecule is formed of two α chains, each of which contains 141 amino acids, and two β chains, each composed of 146 amino acids. The sequence of amino acids is specific in each kind of chain, and variant forms of hemoglobin may differ in only a single amino acid. One well-known variant, sickle hemoglobin, $Hb_{\alpha_2}{}^{A}{}_{\beta_2}{}^{S}$, differs from normal adult hemoglobin only in the substitution of the amino acid valine for the amino acid glutamic acid in the beta chain of normal hemoglobin. Yet, the phenotypic consequences of this single amino acid substitution are extensive. Individuals who are homozygous at this locus suffer from sickle cell anemia; when red blood cells from these individuals are placed under reduced oxygen tension, the red cells assume a characteristic sickle shape. Clinical symptoms of this condition include severe anemia and enlargement of the spleen.

Another kind of hemoglobin, fetal hemoglobin, or $Hb_{\alpha_2}{}^{A}{}_{\gamma_2}{}^{F}$, is normally replaced by adult hemoglobin within a few months after birth, but in a number of medical conditions fetal hemoglobin continues to be produced after this time. In thalassemia major, also known as Cooley's anemia, many red blood cells with a low average hemoglobin content are formed, and both hemoglobin *A* and fetal hemoglobin are present. The defect responsible for the more common form of this disorder seems to reflect partial or complete failure to form normal hemoglobin chains.

Among the over 100 forms of hemoglobin now recognized, most of the adult variants involve a single amino acid substitution. Hemoglobin *C*, for example, involves the substitution of lysine in the beta chain for glutamic acid present in *A* hemoglobin. A large number of variants have been found in human populations in various parts of the world, including the well-known *C*, *S*, and *E* hemoglobins, while many rare variants have so far been found in only one or a few tested individuals or families. Studies of the distributional patterns of several of the more common adult hemoglobin variants have contributed greatly to our understanding of evolutionary processes. The presence of populations in tropical West Africa with 20 to 30 percent of sickle cell trait has been convincingly argued to indicate a selective advantage of heterozygotes for this hemoglobin variant in areas of holoendemic malaria. The presence of hemoglobin *C* in a more limited area of West Africa and of hemoglobin *E* in parts of Southeast Asia may also represent a genetic adaptation to malaria.

The incidence of certain traits other than hemoglobin variants has also been associated with the geographic distribution of falciparum malaria. The frequency of heterozygotes for the thalassemia allele(s), for example, shows a high degree of correlation with the incidence of malaria in the Mediterranean area, while the frequency of an enzyme deficiency, that for glucose-6-phosphate-dehydrogenase (G6PD), also parallels the distribution of falciparum malaria. This enzyme is active in the breakdown of glucose to lactate, a conversion which provides energy for the metabolic activities of red blood cells. Individuals who have this deficiency ordinarily experience no adverse effects, but under certain circumstances (exposure to particular drugs, infective agents, and certain other materials) their red cells may hemolyze, or burst. During the Korean War, many American black soldiers experienced varying degrees of hemolysis when an anti-malarial drug, primaquine, was administered to them; further studies indicate that almost 10 percent of American blacks have reduced activity of the G6PD enzyme.

Another form of G6PD deficiency has been identified in some populations living in the Mediterranean basin area and the Middle East. Pedigree studies reveal that G6PD deficiency is an *X*-linked trait and that its frequency, expectably, is higher in males than in females; but many variant forms of the enzyme deficiency have now been identified. Most of these variants are very rare, but one of the few more common forms predominates in different regions of the world. *Gd(B–)* is the prevalent form of G6PD deficiency in the Mediterranean area, while *GD(A)* forms are more common in African populations, for example.

In addition to polymorphisms of antigenic systems, hemoglobin, and red blood cell enzymes, a number of serum proteins have also been utilized in anthropological studies. Blood, of course, consists of the red blood cells, white blood cells, and platelets suspended in the liquid plasma. More than 90 percent of plasma is water, and about 1 percent of the content of plasma consists of such non-protein substances as salts, nutrients, and waste products. The remainder is composed mainly of a large number of proteins. When one of these proteins, fibrin, is removed from the plasma, the remaining fluid is called serum, which contains, among others, a group of proteins of anthropological interest, the globulins.

The globulins include alpha globulins, formed of carbohydrate and protein; beta globulins, lipids and protein; and gamma globulins, which include most of the circulating antibodies in the blood. One of the alpha globulins is haptoglobin, which can bind free hemoglobin released from destroyed or aging red blood cells to prevent the loss of this vital material from the body by excretion through the kidneys. The haptoglobin molecule is composed of two alpha and two beta chains, but the major variant forms seem to differ due to changes in the alpha chains. Three phenotypes, produced by a pair of codominant alleles, Hp^1 and Hp^2, are identifiable in most tested human populations, although their frequencies vary from group to group. Ahaptoglobinemic individuals, lacking any haptoglobins, are most often found in African populations. When more sensitive tests are applied, two major variants of the haptoglobin 1 phenotype can be identified, called $Hp\alpha^{1F}$ and $Hp\alpha^{1S}$, and the distribution of these variants permits finer distinctions among populations. The faster-moving $Hp\alpha^{1F}$, for example, is rarely found among Australian aborigines, but is common in some African populations.

One of the β globulins which has also been used in anthropological studies, transferrin, also plays a role in the formation of hemoglobin: the transferrins transport ions of oxidized iron, and each heme molecule contains an atom of iron. Twenty or more variants of transferrins are known, but the three major phenotypic forms are *TfC, TfD* and *TfB,* and a number of rare alleles are probably responsible for some of the unusual forms which have now been recognized. Several subtypes of the *TfD* form have a somewhat restricted distribution, with TfD_1 found in African, Australian, and New Guinea populations and TfD_{chi} among certain groups in China and Southeast Asia. Because the transferrins are large molecules, with the protein portion containing over 600 amino acids, it seems likely that additional variants may yet be discovered, particularly since comparisons of several major variants made to date indicate that differences in single amino acid substitutions account for several of the distinguishable forms.

Human populations differ in respect to a rather large number of biochemical traits, of which only a few have been described in this section. It is clearly no exaggeration to say that the study of these attributes has revolutionized physical anthropology, and not only from the standpoint of increasing our awareness of the numerous differences within and among populations. The intimate connections between the nucleic acids and their products are more evident when we can examine the distinctive effects of single amino-acid substitutions on biochemical molecules than when we try to relate genetic differences with traits, such as adult stature, which are far removed from the primary action of the gene and are subject to extensive modification by any of a number of environmental factors.

But it is not enough merely to describe a growing number of biochemical traits and variants in human populations. We want to know how these differences are maintained, why populations differ, and what evolutionary mechanisms or forces underlie changes in the frequencies of the various alleles which direct their expression. Some may, in fact, be traced ultimately to "neutral" mutations, differences which have no selective advantage or disadvantage. Some similarities may reflect ancient genetic relationships. But some of these polymorphisms seem to confer a selective advantage, others are suspected of contributing to the adaptive success of the groups in which they are found, and some differences may represent different genetic responses to similar or common environmental pressures. A good case has already been made for the selective advantage of the sickle hemoglobin heterozygote in the malarial environment, while other hemoglobin variants, the thalassemia allele and G6PD enzyme deficiency seem to confer a selective advantage to other populations in various areas who are exposed to holoendemic malaria. One of the phenotypes of the blood group antigenic system, Duffy, may also offer protection in the form of lower rates of infestation from vivax malaria. Even some of the globulin forms which are related to hemoglobin manufacture may confer resistance to certain infectious agents which vary in their requirements for or their tolerance to different levels of iron.

Our knowledge of these differences and our understanding of the roles of these molecules in the functioning of humans living under a variety of conditions depends on our growing knowledge of biochemistry and of the metabolic pathways by which human life is maintained. Even as this book is written, new discoveries are being made which will further help to unravel some of these problems. We cannot become biochemists or serologists or molecular geneticists, but modern anthropologists have no choice but to develop the ability to communicate with these specialists and to incorporate their findings into our own studies.

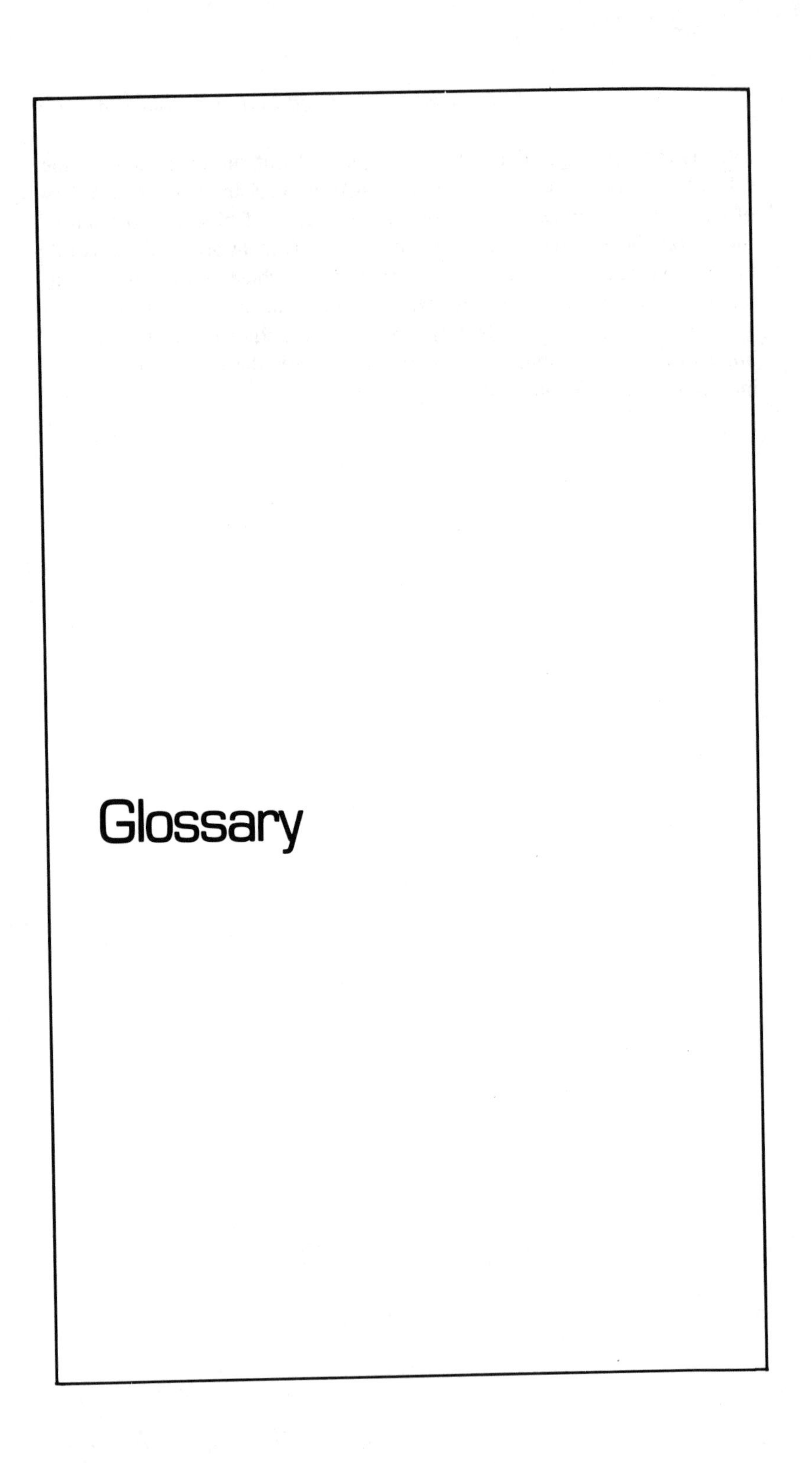

Glossary

Adaptability The capacity of a population to respond to environmental pressures through genetic changes or modifications in physiological, behavioral, or other mechanisms.

Adaptation, Genetic Changes in the genetic constitution which enhance the biological success of the population.

Alleles Alternative forms of a gene at a single locus.

Allen's Rule The ecological principle which predicts that regional populations of mammals living in cold climates will have extremities that are shorter relative to trunk size than those found among related populations living in tropical zones.

Assortative Mating Patterns of non-random mating, including:

Positive phenotypic assortative mating Preferential mating between individuals similar in respect to a phenotypic trait (homogamy);

Negative phenotypic assortative mating Preferential mating between individuals dissimilar in respect to a phenotypic trait (heterogamy);

Genotypic assortative mating Preferential mating between individuals who are genetically related (inbreeding).

Balanced Genetic Polymorphism The continuing presence of two alleles at a locus and at frequencies above those which can explain the less frequent allele as maintained by mutation pressure alone.

Barr Body The structure found near the nuclear membrane in cells of human females; possibly represents an inactivated *X* chromosome (sex chromatin).

Bergman's Rule The ecological principle which predicts that body size of related species of mammals and other warm-blooded animals tends to be larger in colder climates than in warmer habitats.

Blastocyst The structured group of cells produced by mitotic division of the zygote and which gives rise to the organism and the membranes surrounding it.

Chromatids The duplicated portions of the metaphase chromosome joined by a single centromere.

Chromosome Stainable structure in the nucleus which can be visualized under proper treatment in the early stages of mitosis and meiosis; composed of proteins and nucleic acids.

Autosomal chromosomes 22 pairs of chromosomes present in most body cells that are not directly responsible for chromosomal sex determination in humans.

Sex chromosomes The *X* and *Y* chromosomes that direct sexual differentiation in humans.

Homologous chromosomes Chromosomal pair members which are similar in form and structure.

Diploid chromosomal number The total number of chromosomes characteristic of most body cells; $2N = 46$ in humans.

Haploid chromosomal number The total number of chromosomes present in gametes; $N = 23$ in human gametes.

Codominance The condition when both alleles are expressed in the heterozygote.

Coefficient of Inbreeding (F) The probability that both alleles at a locus are identical by virtue of descent from a common ancestor.

Cohorts, Birth A group of individuals who are all born during a specified time period.

Consanguines Individuals who are genetically related.

Conspecifics Individuals belonging to the same species.

Convection The upward movement of hot air.

Deme A localized Mendelian or breeding population.

Deoxyribonucleic Acid (DNA) The genetic material, composed of sugar, phosphate radical, and combinations of nitrogenous bases, linked by weak hydrogen bonds to form a double chain structure, characteristically forming a helical chain.

Directional Selection Selection which increases the frequency of one allele at the expense of other(s).

Dizygotic Twins Twins derived from separate eggs, each fertilized by a different sperm.

Ego The centrum in a pedigree.

Embryo The developing child during the first few months of gestation.

Endogamy The rule or practice requiring a person to marry within the social group of which he or she is a member.

Endothermy Condition in which body temperature is maintained by waste heat of metabolism (homeothermy).

Epistasis When the expression of a trait is determined by the interaction of alleles at more than one locus.

F (see **coefficient of inbreeding**)

Fecundity The biological potential for reproduction.

Fertility Actual reproductive performance.

Founder Effect Form of random genetic drift produced by sampling variance of a small group of founders from a larger ancestral population.

Gamete The mature haploid sex cell, either ovum or spermatozoon.

Gametogenesis The processes of sex-cell formation, oogenesis, and spermatogenesis.

Gene Mimics Different genetic loci capable of producing similar phenotypic expressions.

Gene Pool The sum total of genetic variation potential represented in the genomes of members of a Mendelian population.

Gene Recombination The formation of new combinations of genetic material as a result of crossing over between chromosomal parts during meiosis.

Genome An individual's total genetic component.

Genotype The genetic constitution, either at a single locus or for all loci.

Hemizygous The presence of a single allele at *X*-linked loci in males.

Heterogamy (see **assortative mating**)

Heterosis Enhanced variability and/or fertility in the F_1 progeny of crosses between members of different species, subspecies, races, or populations.

Heterozygous The presence of unlike alleles at a single genetic locus.

Homogamy (see **assortative mating**)

Homozygous The presence of like alleles at a single genetic locus.

Hypoxia Inadequate oxygen in the lungs and blood.

Inbreeding Mating between genetically related individuals.

Incest Mating between socially defined relatives.

Incomplete Penetrance The condition in which a trait is not expressed in an individual who carries the appropriate genotype for its appearance.

Intermixture Exchange of genes between members of different populations.

Klinefelter Syndrome The characteristic set of features, including small male type genitalia, long limbs, and sparse body hair, expressed by individuals possessing a *Y* chromosome in combination with two or more *X* chromosomes.

Lethal Alleles Alleles which do not permit survival of the embryo or infant.

Leukocytes White blood cells.

Levirate Marriage of a woman to her deceased husband's brother.

Load, Genetic The proportion by which fitness is reduced in a population due to the operation of factors such as mutation, segregation, or incompatibility interactions.

Locus The site on a chromosome occupied by a particular set of alleles.

Meiosis The process of reduction-division leading to the formation of haploid nuclei.

Menarche The beginning of menstruation.

Migration Effect Commonly observed differences between migrants and their offspring in such features as height, head form, or bodily dimensions.

Mitosis The process of cell division.

Monozygotic Twins Twins derived from the early division of a single fertilized egg.

Mutagen Any agent (ionizing radiation, certain chemicals) capable of inducing a mutation.

Mutation Any change in the genetic material other than changes due to gene recombination.

Neonate The newborn infant, from the time of birth to age 28 days.

Oocyte The egg cell before it is fertilized.

Ovum The mature female gamete.

Panmixia Random mating in respect to any trait being studied.

Phenotype Observable or measurable characteristic(s).

Plasticity Ability to respond phenotypically to environmental stress.

Pleiotropic Effects Multiple effects or expressions of a single genetic locus.

Polygenic Inheritance The inheritance of traits whose expression depends on two or more genetic loci with alleles at each locus contributing small increments to the total expression of the trait.

Polygyny The marriage of a man simultaneously to two or more wives.

Polymorphism The presence of two or more variants of a trait among members of a single population.

Polypeptides Chains of amino acids.

Polyploidy The presence of additional whole chromosomal sets (e.g., in humans, the triploid number is 96).

Population

Stationary population A population neither increasing nor decreasing in size.

Stable population A population with constant age-specific vital rates which may be increasing or decreasing in size.

Population pyramid Diagrammatic representation of the age-sex distribution of a population at a certain time.

Polytypic The presence of differentiated populations within a single species.

Rad A unit of the absorbed dose of radiation.

Roentgen A unit of ionizing radiation, slightly smaller than a rad.

Sex Chromatin (see **Barr body**)

Sex Influenced Inheritance Refers to autosomally transmitted traits, the expression of which is affected by hormonal levels.

Sex Linked Inheritance Refers to traits determined by genetic loci present on the sex chromosomes.

Sexual Selection Differential access of some males with economic, political, or social advantages to females.

Sex Ratio The ratio of males to females.

Primary sex ratio Ratio of males to females at the moment of conception.

Secondary sex ratio Ratio of males to females at birth.

Siblings Individuals related by virtue of descent from a common set of parents.

Sororate The marriage of a woman to her sister's husband.

Spermatocytes Immature male gametes.

Spermatozoon The mature male gamete.

Stabilizing Selection The form of natural selection which operates to maintain the status quo, usually by removal of extreme forms.

Steatopygia The presence of a projective fat deposit in the buttocks area.

Transient Genetic Polymorphism The presence of two or more alleles at a locus in a population undergoing directional selection that has not been completely effective.

Turner's Syndrome The characteristic set of phenotypic features (short stature, undeveloped breasts, webbed neck, etc.) in individuals with the *XO* chromosomal complement.

Variable Expressivity Variable degrees of phenotypic expression for certain traits in individuals with identical genotypes at this locus.

Vasoconstriction Constriction of blood vessels, particularly in the extremities.

Vasodilatation Dilation of blood vessels.

Vital Rates

Crude vital rate A demographic index expressing the ratio between the number of events (births, deaths, migration) in relation to the entire midyear population.

Age-specific vital rate A demographic index expressing the ratio between the number of events (births, deaths, etc.) in relation to the number of individuals belonging to specific age-sex cohorts.

***W* (Darwinian Fitness)** A measure of the relative fitness of one genotype in comparison to other genotypes at a specific locus.

Zygote The fertilized egg cell.

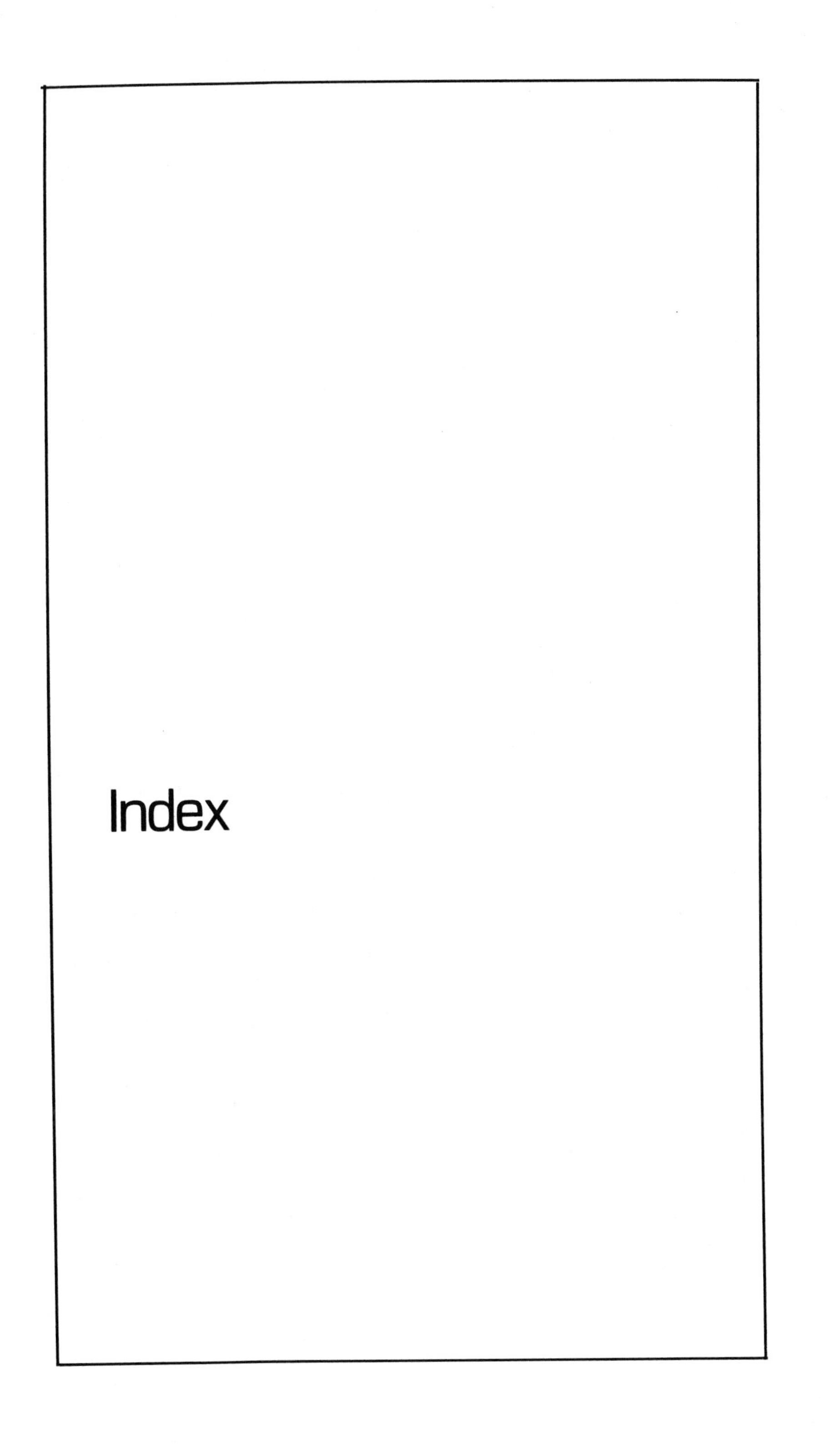

Index